T0073859

Scent

SCENT

A NATURAL HISTORY
OF FRAGRANCE

ELISE VERNON
PEARLSTINE

ILLUSTRATIONS BY
LARA CALL GASTINGER

Yale
UNIVERSITY PRESS

New Haven and London

Published with assistance from the foundation established in memory of
Amasa Stone Mather of the Class of 1907, Yale College.

Yale University Press books may be purchased in quantity for educational,
business, or promotional use. For information, please e-mail sales.press@
yale.edu (U.S. office) or sales@yaleup.co.uk (U.K. office).

Set in Adobe Garamond type by IDS Infotech Ltd.
Printed in the United States of America.

Library of Congress Control Number: 2021948038
ISBN 978-0-300-24696-4 (hardcover : alk. paper)

A catalogue record for this book is available from the British Library.

This paper meets the requirements of ANSI/NISO Z39.48-1992
(Permanence of Paper).

10 9 8 7 6 5 4 3 2 1

To Leonard, who listens to my stories, and to Alison, Mike, Ben, Tim, Andrew, and Avery

To Michelyn, who helped me find my voice

Contents

CONTENTS

Preface

Before history recorded such things, people would strew aromatic herbs on a floor, use pine boughs to freshen a home, or blend flower petals in oil to rub on chapped hands as a way to add fragrance to their lives. Today there may be an aromatic candle or two, fragrant lotion, perfumes and potions, and even scented toilet paper. Gardeners value aromatic plants, waiting impatiently for their fragrant lilacs to bloom, and nearly everyone will bury their nose in a bouquet of flowers to get a sniff of the floral and green. Visit any bookstore and you will find books about gardening, even gardening for specific interests or regions. There are also an increasing number of books on the behavior of plants—how they respond to each other, attract pollinators, and interact with their environment. You can find blogs and books about perfumes and fragrance. What ties these themes together are aromatic chemicals called plant secondary compounds that play a role in the protection, pollination, behavior, environmental response, and health of plants. Blending story and science, this book will take you through history and around the world to investigate the smelly molecules that plants produce and how they have affected our world. Although we love our fragrant plants and their products, humans have played a minor role in the evolution of aromatic compounds, also called volatile organic compounds, or VOCs. It is the moth, the bee, the beetle, the bat, the fungus, and the microbe that have driven evolution in plants to give us ingredients for perfume, incense, potions, and medicine. As a perfumer and naturalist, I have found the inspiration to research and write this book through my appreciation for scented plants and their tiny friends and enemies.

The natural history of fragrance involves tales of incense, spice, gardens, and perfume. From frankincense to jasmine, plants show us their relationship with humans and the world around them through the functions that aromatic chemicals play in flower, leaf, and seed. Environmental characteristics, often called terroir, influence the volatiles that a plant produces, and this book places familiar scented plants and their products in the soil of their homeland. History has turned on aromatic products such as incense and spice, and industries have been built on fragrant ingredients. Ancient and modern global patterns of trade and manufacturing provide context and contrast with the intimate experience of smelling a flower from your garden, cooking with spices, or perfuming your personal spaces.

Fragrance perception is a difficult thing to quantify, very personal, and one of the senses for which we have few descriptors. For the plants in this book, I have included descriptions of scent that are derived from my knowledge of plant essential oils and extracts gained from years of comparison and contrast to build my scent memory, as well as from descriptions that can be found in the literature. Not everyone smells things the same way, and my goal is to provide some groundwork on how to experience and describe a scent and to show you a way to smell, describe, and enjoy the fragrant world that surrounds us all. Writing about volatile organic chemicals involves naming and describing the molecules that influence a plant's interactions with the world. Depending on your interest you may want to treat the constituents in the book as tools the plant uses, or you may be interested enough to read more about the chemistry behind fragrances. Using plants as medicine is a practice as old as time and the subject of much current research, but aside from a few instances, I have chosen to limit discussion of medical uses to keep the focus on fragrance.

Although we did not cause the rose to breathe out scent or the frankincense tree to shed fragrant tears, our stories are intertwined and mediated through the plant secondary compounds that we call fragrance. Released from damaged leaves, delicate petals, resinous trunks, green stems, pollen, nectar, seeds, and leaves, plant volatiles are made in response to pollinators, predators, and pathogens. Apart from a small percentage of plants bred for scent, the smells we appreciate, and use, are created neither by us nor for us.

To tell the story, my tale will follow the history of our use of aromatics from ancient times to modern day with themes of spirituality and mysticism; power, revolutions, and control; gardens; and fragrance as perfume, industry, and fashion. For people, the story begins with mysticism and with scented smoke rising to the heavens, taking its sweet smell as a message to the gods. Resins such as frankincense, myrrh, cannabis, and copal and woods such as sandalwood and agarwood have a long history of use for incense and trade. Molecules from the wood of trees are complex; they are often large, they are healing, and they are greatly valued in perfumes, traditional medicine, and religion. Resins smell of pine, lemon, freshness, and terpenes, smaller molecules that can be experienced in woodland walks or the rising smoke of incense.

Small enough to fit into little bottles on a kitchen shelf and yet with a huge fragrance, spices have had a worldwide influence on trade and exploration. There was a time when the mysterious sources of such spices as pepper, clove, nutmeg, ginger, and cardamom were the subject of secrecy and legends. Explorers followed those elusive hints to find the sources and help establish empires and create vast wealth. Spices are the flowers, fruit, bark, roots, and seeds of herbs, vines, and trees, each producing characteristic fragrances and tastes from a suite of molecules that are often antimicrobial and protective in nature. Their habitats are the warm woodlands and remote islands of the world.

Gardens are a home for the herbs, roses, and flowers that make perfume for their pollinators, but their perfume also attracts us so that we surround ourselves with their beauty and fragrance and tend them carefully. People, it seems, have always had gardens, and gardens have always been as varied as the people who planted them. Herb gardens are often simple in their construction, and their purpose is straightforward—to grow lavender, rosemary, and other scented plants for healing and cooking. Wealth, on the other hand, provides for grand gardens and rich displays of plants and patterns—providing a place to retreat, "smell the flowers," and exclude the outside world. Plant hunters have always sought out the unusual, some of which are now found in botanical gardens around the world. That iconic garden flower, the rose, has a long and complicated history, and some thrive outside our gardens in the wild and around old buildings.

A book about fragrance would not be complete without the story of perfumery and the extraction of scented compounds for the purpose of beauty and attraction. Flowers are masters of creating and blending fragrant molecules and may produce hundreds of individual aromatic compounds that together create a unique floral scent—a scent that reaches out to moth, butterfly, bee, or beetle, enticing them to come near, sip a little sweet nectar, and oh, by the way, deliver this little packet of pollen to that other plant over there. The relationship between moth and flowering tobacco provides an intimate example of attraction, repulsion, and protection based on these volatile molecules. In the south of France, a city called Grasse grew in the limestone mountains near the blue Mediterranean Sea where lavender was found and jasmine thrived. Perfumes constructed of fresh citrus, floral hearts, and musky woody bases were the foundation of the early business of perfumery and aromatic extraction.

Modern perfumes were born at the junction of industry and science where synthetic molecules were created in the lab. Perfumers began using them to create fantasies, and visionaries such as Coco Chanel saw the opportunity to pair fragrance with fashion. Single molecules and formulated blends took the place of the more expensive plant extractions to give the world the hundreds, if not thousands, of perfume launches seen yearly in today's fashion world. New tools and consumer demand have led fragrance companies back into the field to analyze plants and to use little microbial factories as biological producers of fragrance.

The world is at a turning point as I write this. We are in the midst of climate catastrophes and (I cannot believe I am writing this) trying to figure our way out of a global pandemic. As you read this book your attention may be caught by stories of the many small farms and indigenous people who grow your spices or the iconic and endangered conifers that people love enough to give names like The Senator or Gran Abuelo. You may notice the tenuous balance between temperature and precipitation in the Mediterranean, where favorite herbs and essential oil plants are grown, or the challenges facing forested landscapes that produce the mysterious agarwood. Knowledge can be power, and I hope that readers will use knowledge gained from this book to help in making fragrant choices. It is also my fervent hope that readers will take the opportunity to slow down and appreciate their aromatic surroundings, to pause in their walks for just a moment under a tree or near a flower bed, to plant a few scented flowers or herbs in their gardens, or even learn a new recipe or two using unusual spices. Then go out and support a local botanical garden, help in park cleanup activities, check into using organic gardening methods, and find a path to sustaining the soils, insects, microbes, and plants of this world.

This is not simply a book about perfumes and perfumery but one about where fragrant plants fit in our history and our lives. It is a book

for gardeners, perfume aficionados, fragrance lovers, hikers and walk-
ers, cooks, people who love insects, homeowners who have killed their
lawn to grow pollinator plants, and people who sniff everything, are
passionate about roses or lavender, or want to know more about natural incense. Perhaps you simply think that plants and insects may have
a secret life, and you want to know more about that world.

Acknowledgments

Thanks to my editor, Jean E. Thomson Black, for believing in a book about fragrant molecules and for her skillful guidance; Martha Hopkins, my extraordinary agent, who has guided me on this journey and who calls when she sees bees and flowers; and Alison Larsen for helping me figure out how to tell this story. To Leonard Pearlstine for his support over this long journey, for being my sounding board, and for getting me started on my fragrant path. To the talented team at Yale University Press: Mary Pasti as production editor; Amanda Gerstenfeld and Elizabeth Sylvia as editorial assistants; book designer Dustin Kilgore; and production manager Katie Golden. To Laura Jones Dooley as manuscript editor. To Anya McCoy, teacher and mentor, and Mandy Aftel, teacher and inspiration. To Jessica Hannah, Maggie Mahboubian, and Melanie Camp for getting me started and helping me think I could do this. To Dr. Giby Kuriakose for sharing his experiences with cardamom in the field, and to Trygve Harris for sharing her on-the-ground knowledge of frankincense. To Donna Hathaway, Susan Marynowski, Bonnie Kerr, and Ida Meister for reading various drafts and offering helpful insights. To Dr. David Howes, Dr. Stephen Buchmann, and an anonymous reader for insightful comments and support. To Michelyn Camen and the team at ÇaFleureBon for exquisite writing about fragrance and giving me a start. To Eric Larsen for telling me all about the technique of underpainting and understanding how it relates to jasmines. To Douglas

Decker, Katlyn Breene, JK DeLapp, Dan Riegler, artisan perfumers, and incense-makers for helping to inform my fragrant world. To Christopher McMahon for his knowledgeable blog posts about exotic aromatics and for many wonderful oils through the years. To Fern and Leo Vernon—Mom and Dad—for always having a garden: Dad for moving giant Utah boulders with a five-foot-long crowbar to create structure, and Mom for making it pretty. To my siblings, Rick, Marty, Jill, and Eric, for sharing in the adventures and helping me remember stuff. To plants for taking sunshine, air, water, and dirt to create beauty and scent. And to our planet and those who care for it garden by garden, park by park, and wildland by wildland.

Introduction

Hold the leaf of a lime tree up to the sun to see hundreds of tiny pockets of scent, or peel the fruit for a little spritz of freshness. Take a hike in the woods and put your hand on a large pine tree—your hand may come away with sticky resin smelling of woods and sunshine. Place scented roses in a vase to enjoy the aroma that fills the room and perhaps remind yourself of a loved one. Grind tiny black peppercorns to appreciate their pungent fragrance of spice, wood, and citrus. Then take a moment to appreciate the plants that have created these fragrances, not for us but for moths and beetles, bacteria and fungi, bees and flies, pollinators and predators. This is the story of how and why plants create and manipulate volatile compounds as they interact with the world around them to attract pollinators, fight disease, turn away herbivores, and heal themselves. It is also a story of people and their plants from around the world that will follow the arc of history and culture from prehistory through the Middle Ages to the industrialized world. On this journey we will find smoke, faith, secrecy, power, nation-building, wealth, addiction, repulsion, fashion, and seduction.

For me, as with many of us, the story of fragrance begins with my immediate environment—the smell of cookies baking, the comfort of a scented candle, rain falling on creosote after a dry desert summer, a pretty pink orchid, and the smell of soil in a fertile garden. My fascination with aroma is intricately tied up with my work as a perfumer: I draw on my experiences and the stories of fragrance to help me make my perfumes, to re-create in some way what plants do without thought. I sometimes start by staring at my little amber bottles full of aromatics, each as familiar to me as my fingers, and call on my scent memories as I create fragrance. At other times I dip a narrow paper blotter into one amber bottle, close my eyes, and breathe in. Then I add another blotter with a different aromatic so I have two together—building a fragrance piece by piece and breath by breath as I inhale scent and make smell the only thing that matters. Sometimes the ingredients conspire to tell a story, evoke an emotion, or perhaps just fulfill the yen for creativity through perfume.

Plants are mostly familiar to us as food—fruits, vegetables, and grains that sustain life—but we also value and use them for fragrance. Although we did not cause the rose to breathe out scent or the frankincense tree to shed fragrant tears, our stories are intertwined and are mediated through what are called plant secondary compounds. In order to live, plants require food and a living structure, to which end they make proteins, fats, and carbohydrates for support and nutrition. Because these primary compounds are necessary for essential life functions in the metabolism of a plant, this is where a plant focuses most of its energy and resources. These molecules may also be manipulated to form scented compounds, one example of which is the transformation of colorful carotenoids into violet-scented molecules called ionones. Secondary to growth and nutrition are other critical life processes such as reproduction and disease resistance. Plants are unable to relocate to find a mate or avoid disease, and so they release fragrant chemicals into the air from flowers to attract pollinators or volatile molecules from damaged leaves and stems to deter predators, resist disease, and heal tissues. This really means that it is the moth pollinators, the little insect predators, and the bacterial and fungal diseases, not humans, that influence fragrances in plants. With the exception of a small percentage of plants bred for scent, the smells we appreciate and use are created neither by us nor for us. We have, however, valued the scent and healing properties of plants since humans first chewed on a mint leaf or used branches from an aromatic pine tree to make fire.

INCENSE, WOOD, AND RESIN

For humans, the story begins with mysticism and with scented smoke rising to the heavens. Fire once meant protection against the dark and the scary, but it also transformed wood and resin into the ephemeral, creating an uplifting aroma that ascended to the night skies. Frankincense and myrrh, resin-producing trees, have a history going back to the early Egyptians, who used resin from both in their temples for worship and to preserve the dead. Tons of resinous tears from frankincense and myrrh traveled by boat from their origins on the Horn of Africa and aboard camel caravans from deep in the hostile deserts of the Arabian Peninsula to the rest of the world via early trade routes connecting East and West. The trees, growing slowly in dry and rocky terrain, produce scented resin as a balm to coat injured bark and protect themselves from pathogens looking to infect. On the other side of the world, trees of the Americas produce copal resins that have long been used in worship and mysticism. Age produces beauty in the heartwood of sandalwood trees as essential oil concentrates in the oldest branches, trunk, and roots, darkening the wood over the years and formulating a rich and precious aroma. To extract the fragrant oil the tree must die and the wood must be distilled, often through techniques and tools that have remained unchanged for centuries. Another valued tree, agarwood, from southern Asia sometimes develops a darkly resinous heartwood because of infection. Also known as oud or eaglewood, its fragrance is not easy to love, with a hint of barnyard but also notes of tobacco and leather or even berries. And yet agarwood is one of the most sought-after and valued of aromatic wood products while being loved nearly to death—almost all naturally growing trees have been felled.

Frankincense tree (Boswellia sacra) *from Oman*

1

The Torchwoods:

Frankincense, Myrrh, and Copal

Small and gnarled, often showing its age, the frankincense tree grows in an Arabian Desert wadi where water occasionally flows in the winter. Its bark is papery and peeling in places, on it a tiny beetle is trapped in resin, and the trunk glistens where it is coated with the sticky stuff. Farther below there are small drops like liquid tears that turned solid as they traced a path down the trunk. If you were to put a hand on the tree, you might come away with a bit of scented resin softened by the hot sun into a sticky patch: the scent is resinous but also lemony and soft and subtly appealing. Resins are produced and secreted in specialized structures on the surface of a plant or within its tissues; sap, by contrast, circulates throughout a plant carrying water and nutrients.[1] Resins are also characterized by the presence of volatile chemicals—the most abundant are called terpenes—and may be important to a plant's interactions with the world around it. Generally produced within the trunk of a tree, resins may also exude from leaves, new growth, cones, or flowers in a sticky, somewhat liquid form, and they vary in hardness and such other physical qualities as stickiness, color, and fragrance. Many plant groups have evolved to produce resins that may have similar constituents but are blended differently to serve a variety of purposes, from defense against herbivores and disease to providing a layer of protection against desiccation and damaging ultraviolet light. Conversely, resins may attract pollinators and other animals that use the sticky substance for protection, nest building, or shelter.

Frankincense and myrrh trees both exude resin from their trunks in the form of tears. Frankincense tears come in colors ranging from nearly white to a lovely green to a dark amber color, while myrrh resin is almost always a transparent brown. The fragrance of frankincense is resinous, with hints of citrus or flowers, while myrrh is more bitterly medicinal but with a mysterious richness. On the other side of the world, close relatives of frankincense grow in the tropical forests and deserts of Central and North America to produce a similarly scented fragrant resin called copal. Trees are not the only plants that produce resins; sometimes a flower produces the sticky stuff, as is the case with the cannabis plant. The stinky, glistening flower buds of cannabis are highly resinous and filled with aromatic chemicals and psychoactive constituents. Hemp, from the same plant, is also highly useful in fiber production and is likely one of the earliest cultivated plants: its buds and seeds have made their way around the world in the centuries since its discovery by humans.[2]

As we learn about fragrance, the how and the why, in these pages, we will explore their fragrant constituents with chemical names. These are important because they determine both the response of the plant to environmental challenges and the fragrance that brings value to human society. In resin-producing plants, fragrance is all about the terpenes. Terpenes, the most diverse class of plant compounds, are the smoke of the Smoky Mountains, the fragrance of a Christmas tree, the freshness of a citrus peel, the highly varied aromas of cannabis, the turp in turpentine, and the sharp and floral notes in spices. Monoterpenes are organic molecules organized around a backbone with ten carbons, making them small in the world of organic compounds. This volatility makes them useful as both the top notes of perfumes and the pungent notes of spices that come to our nose first as we spritz or grind them. Sometimes an additional five-carbon chunk is added to the monoterpene to create a sesquiterpene molecule. Being larger, sesquiterpenes are somewhat less volatile, meaning they evaporate more slowly—they are important fragrance elements in sandalwood and agarwood and comprise some of the more subtle components of other woods and spices.[3]

Terpenoids are regular constituents of resin in plants and are a means by which plants interact with the world around them. Often, the name of the molecule hints at its fragrance. Pinene is found in conifers like pine trees but also in spices such as black pepper, cardamom, and allspice; in herbs such as basil, dill, lavender, rosemary; and fruits such as blackberry and citrus. Limonene is pretty much the pure smell of citrus and, to some people, the smell of household cleaners. A huge presence in most citruses, limonene is also found in many spices, fruits, and evergreen trees. Myrcene is also found in spices, citruses, eucalyptus, and herbs. It has a fragrance that can be described as spicy or herbal, balsamic (meaning resinous but also sweet vanilla in scent), and even slightly rosy. There are many more terpenes, but these few show up often and across a variety of plants and plant parts where they may repel herbivores or attract predators that eat the herbivores but will also blend into the fragrance of a flower. Sesquiterpenes are slower to evaporate and arrive at our nose and provide the complex scent of woody fragrances. The highly complex fragrance of resinous agarwood is a product of more than 150 aromatics, some being sesquiterpenes, and sandalwood develops its scent, not as a resin, but as heartwood that accumulates sesquiterpenes, giving it a charismatic, precious wood fragrance that is buttery and yet animalic.

Here we begin the story of volatile molecules and the plants that produce them—the where and the why of plant aromatics—and the diversity of plant mechanisms, life stories, habitat, and relationships behind the scent. Because this story is also about people and the influence of fragrance on our history and civilization, there will be tales of humans, roads, religion, ritual, smoke, perfume, and incense in narratives that are as old as time and as wide as the world.

Frankincense trees in the genus *Boswellia* grow in the coastal areas of the Arabian Peninsula and are the most famous members of the Burseraceae, or torchwood, family. Sparse forests of these trees grow among rocks and sand at the southern tip of the Arabian Peninsula where monsoon-watered mountains meet dry desert habitats, and they make do with very little. The gift of scent these ancient trees return from their stark habitat is highly sought after and held sacred by many. Highly valued throughout history, frankincense as a trade good was tied to domestication of the camel and development of the Incense Road through the Arabian Peninsula, which brought currency, goods, and progress to the region as early as 1500 BC.[4] The name frankincense reminds us that it is the very definition of incense; it derives from the old French *franc encens,* which means pure incense or pure lighting. Aromatic compounds in the resin are produced within the plant's tissues—protective in nature, they help resist infection by fungus, repel attacks by insects, prevent desiccation, and seal injured tissues. The resin of frankincense is generally light colored and leaks as tears that flow for a bit and then harden and solidify around wounds in the bark of the tree.

When you first experience frankincense tears, a typical resin odor may overwhelm the other constituents, but if you can get your nose to bypass those fragrances, you can perceive a variety of fresh, citrusy, and even floral notes. The best way I have found to get the full effect of the fragrance is to gently warm the tears so they melt and perhaps smoke a bit, but not to the point of burning. Hojari or Omani frankincense, from *Boswellia sacra* in Dhofar, is a lovely light green or lemon-colored resin. These high-quality tears have a classic frankin-cense woody-resinous fragrance but also a fresh sweetness and earthi-ness layered on top of amber. One of my favorites is maydi, or *B. frereana,* from Somalia. It is golden amber in color and has a pro-nounced fresh and lemony fragrance interwoven with the typical frankincense aroma. Another species, *B. papyrifera,* comes from So-malia and the Horn of Africa and is used by the Catholic church in its incense.

Another resin, similar to frankincense, is myrrh. There are many species in the genus *Commiphora,* but it is *C. myrrha* that is generally used in medicine and to produce essential oils. Myrrh trees grow in the same habitats as frankincense, also produce a healing resin, and were a valuable trade item along the Incense Route. The dark, amber-colored resin may be used in the same blends with frankincense and is an im-portant ingredient in incense, perfumery, and traditional medicines. Myrrh shares much of its history, ecology, and uses with frankincense.[5] The fragrance of myrrh essential oil is warm, spicy, somewhat balsamic, and medicinal. For a unique and amber touch, myrrh is lovely in per-fumes, but its primary uses are as a companion for frankincense in in-cense and as a traditional medicine for the skin and mouth.

Frankincense trees may look scruffy, but their flowers are sweet smelling, and the resin that weeps from the trunk and stems has been deemed more precious than gold. They grow naturally along the coast of the Arabian Peninsula and in sub-Saharan Africa. To find them, you would have to hike the wadis of the Arabian Peninsula and scramble out onto the dry and rocky limestone cliffs and hills that surround the mostly dry watercourses. Across the waters of the Red Sea and Gulf of Aden, in the poor soil and steep hillsides of the Horn of Africa, you would clamber along another rock-strewn land-scape to find similar sparse dry forests where frankincense and myrrh trees establish themselves. These hardy species take nutrients from the soil and combine them with scarce rain and desert breezes to create something precious and healing—tears for our rituals and support for local communities who harvest the resin for the rest of the world.[6]

Eritrean and Ethiopian frankincense trees, including *Boswellia papyrifera,* grow to around thirty feet tall. These are the resinous trees sought out by ancient Egyptians, and the area around Eritrea is thought to be the mysterious Land of Punt that was their source. In these harsh environments of rocky soils and little rainfall, the trees produce the highest quantities of aromatic resin after they are about eight years old. Frankincense leaves are green and curled, emerging in the spring as clusters at the ends of branches. The bark is papery and thin, peeling from the multistemmed trunk anchored by a swollen base that secures the tree to steep cliffs and rocky slopes. Leaves fall at the beginning of the dry season, and sweet-smelling flowers bloom just before new green leaves bud out in the early spring. Native pollinators love the trees, and local honey bees will visit the flowers for pollen and nectar, producing good-quality honey. Trees seem to do best in harsh, rocky habitats—in good conditions, if left alone and protected from fire and grazing livestock, they seed profusely and regenerate well. However, overuse and other pressures negatively affect natural populations. Goats and camels graze on them, longhorn beetles bore into their trunks, local inhabitants clear the land for agriculture, and resin collectors sometimes overharvest, all to the detriment of the trees. As frankincense has grown in popularity and harvesting increases, scientists are raising concerns about sustainability. In some places trees are not maintaining healthy populations as a result of these pressures. But some studies also show the potential for supporting populations of frankincense by establishing cattle enclosures and fire breaks and tapping the trees sustainably. The resin is an important source of income for many local harvesters by way of foreign cash for countries producing frankincense (and myrrh).[7]

Across the Gulf of Aden, along the coast of Yemen and Oman, *Boswellia sacra* grows just inland from rich coastal cloud forests of the southern Arabian Peninsula. To understand its distribution, we begin our investigation of vegetation and rainfall patterns at the coast and move inland. We learn that between March and October a monsoon wind comes off the Indian Ocean and blows over the peninsula. This sea breeze cools as it rises, and when night falls it brings high levels of humidity, giving rise to dense fog banks that hover just above the higher ground. Moisture leaves the air, either condensing on vegetation to drop slowly to the ground and thirsty roots or gathering to form a kind of fog precipitation. Frankincense trees grow in the moist areas but produce low-quality tears. Farther inland, drier hilly regions are interspersed with wadis. Behind the cool, moist highland, beyond the reach of the precipitation-providing fogs, but where cool winds sometimes blow, are the wadis and dry forests that support the most sought-after frankincense trees.[8]

The town of Salalah in the Dhofar region of Oman will serve as a reference point for a bit of geography. Salalah is located nearly in the center of the southern end of the Arabian Peninsula. A healthy population of *B. sacra* grows in an area called Al-Mughsayl to the west and in the Wadi Dawkah area to the north, both dry scrubland habitats backed by the green mountains of the coast. Frankincense trees here have a lovely branching architecture with umbrellalike crowns and are happiest growing along lonely stone outcrops and clinging tightly to solid rock among periodically flooded valleys called wadis.[9] Wadi Dawkah is protected as a natural park and included by UNESCO among Oman's World Heritage sites as the Land of Frankincense. Throughout the region frankincense trees are an important component of the semidesert areas and, as an integral part of the local ecology, could be called a keystone species. Populations are declining here as in Africa possibly due to similar challenges such as grazing by goats and dromedaries and cutting for firewood. In addition, temperatures are rising and rainfall is decreasing due to climate change. Because the trees in this region grow in an area tenuously balanced between wet and dry, small changes in temperature and rainfall may narrow or eliminate these habitats.

Once harvested, the aroma of frankincense resin is best released with heat and fire and may have reached the noses of early humans via a simple fire for heat or cooking. It is easy to imagine a small group gathered around a fire constructed of frankincense wood for companionship and safety. The fragrant smoke would only have enhanced those emotions. Anyone who has enjoyed a campfire knows that smoke lingers on the body, hair, and clothing of those sitting around the fire. Which, if the smoke is pleasantly aromatic, may have been a good thing in the days before regular bathing. Before long, I would imagine, the precious tears would have been gathered separately to be used for their fragrance. The word *incense* often refers specifically to frankincense but may also be used for other fragrant woods and resins or for specific blends that may include spices, dried leaves, flowers, roots, and other aromatic materials like ambergris that release a pleasant scent when burned or heated. From its first use humans seem to have recognized the rising of scent from incense, both pleasant and evocative, as a prayer or communication with gods thought to be living on high. Incense could carry messages upward, could purify holy spaces, and could give people a focus for meditation and ceremony.

Egyptians were famous users of fragrant materials, whether to scent and preserve dead bodies as mummies, to use as a perfume in the form of an unguent (ointment or salve often made with oils or fats), or to burn as incense.[10] Frankincense and myrrh were common ingredients in the Egyptians' aromatic tool chest, and both were used to preserve mummies. Ancient Egyptians mainly used oils or fats as the base for their perfumes: oils for liquid and solid fats to make unguents. The most listed ingredients in unguents and perfumes include frankincense and myrrh, cinnamon and cardamom, iris and lily, mint and juniper, and other ingredients both locally available and imported. In a process referred to as maceration, chosen ingredients were added to fats in a particular order and for specific lengths of time to control the strength of the fragrances. The Ptolemaic temple of Edfu has extensive records on the walls of what may have been a perfume laboratory (or perhaps a storage room) that seem to be recipes for the commercial preparation of scented materials. Perfumed oils were used to anoint deities, given as gifts, and left in the tombs of royalty. The wealthy and royal also used solid perfumes stored in elaborate containers and were sometimes illustrated as having cones of fragrant unguents on their heads that would melt with body heat, releasing the fragrance.

Made from the pure resin or sometimes blended with other fragrant ingredients, incense in Egypt was burned ritually according to prescribed schedules—frankincense in the morning, myrrh at noon, and the sacred blend called *kyphi* at night. There are many recipes for making kyphi, but they all seem to share raisins and wine, fine resins such as frankincense and myrrh, spices, grasses, conifers, and mastic bound together by ritual. The perfume of the incense connected kings with the realm of gods. In a crossover between religion, meditation, and medicine, kyphi was used as a cure for snakebites (possibly in a potion) and as an antiseptic, an aid to vivid dreaming, and a sacred component of holy ritual.

Trade in the Old World had its cradle in the land of Mesopotamia as long ago as 3000 BC. From their home between the Tigris and Euphrates Rivers, Mesopotamians could exchange goods with their neighbors by land and sea. We can be fairly sure that by 1500 BC a series of caravan routes through the hostile Arabian Peninsula brought incense from the south, where frankincense grew, to intersect with routes of traders that passed through the Fertile Crescent and eastern Mediterranean. As time went on, boats plied the Persian Gulf and the Red Sea to exchange goods from China, the Near and Far East, and southern Arabia to Mesopotamia and Egypt. Silk from China and horses from the nomads of Mongolia were early trade items as Far Eastern traders forged paths between mountain and desert to establish the routes that came to be called the Silk Road. At the other end, land routes from the Mediterranean began as a network of roads from Anatolia extending as much as sixteen hundred miles to connect with the Silk Road.

By around 200 BC, Arabians were bringing incense up from the south and Chinese were traveling the Silk Road. Greeks were sailing through the Red Sea and across the Arabian Ocean toward India by hugging the coast. In about 100 BC a Greek sailor discovered (or rediscovered something African and Indian sailors already knew) a faster way to India and its spices: the pattern of the monsoon winds that pushed sailing ships eastward toward the Malabar Coast in the summer and back to the southwest in the winter. With Greek sailors stitching together the sea routes, ports were developed, routes were refined, and countries from Sri Lanka to Yemen and Persia to Africa moved spices, incense, and silk by sea and land on the Incense and Silk Roads.[11]

In their ongoing quest for fragrant materials, early Egyptians traded in frankincense and myrrh. Queen Hatshepsut, who reigned during the fifteenth century BC as one of only three female pharaohs, organized an expedition to the fabled Land of Punt, possibly located in the area of Eritrea and Somalia, in search of resins. Afterward she ordered a temple built in Deir el-Bahari, on the west bank of the Nile, opposite Luxor, as part of her legacy. Hatshepsut's temple is considered one of the wonders of ancient Egypt, and the beautifully detailed bas-reliefs of the expedition are the finest in the temple. In them, you can see huts on stilts surrounded by date palms and fragrant trees that may be frankincense or myrrh. The hieroglyphics are so detailed that scientists have identified fish in the streams, and you can see goods such as precious resin, myrrh saplings, animals, and gold. It was not long before the Land of Punt became a mythical place, no longer visited and its location a deep secret.[12]

Humans' love affair with frankincense began long ago. Ancient inhabitants of the Arabian Peninsula and the Horn of Africa certainly recognized the beauty and sacred nature of tears of the frankincense tree and incorporated them into medicine, life, and ritual. Frankincense and myrrh made their way to the outside world early in history and formed the foundation of the Incense Road. Once traders discovered frankincense and myrrh, they made every effort to bring them out from the harsh deserts of the Arabian Peninsula. The domestication of the camel in 1500–1200 BC allowed the incense trade to grow. Due to their adaptability to the extreme conditions of the deserts of the Arabian Peninsula, camels were the perfect desert bearers of heavy loads of fragrant incense, progressing from oasis to oasis and assisting in the creation of inland kingdoms.[13] By 1000 BC frankincense was known and valued in Babylon, Egypt, Rome, Greece, and China. The movement of frankincense along the Incense Road, especially at its height between 300 BC and AD 200, was one of the most important trading activities of the ancient world and was responsible for the construction of cities, forts, and irrigation systems in the harsh desert. Because of the high demand for frankincense, kingdoms of southern Arabia connected with India, the Mediterranean, and the Silk Road. By the second century AD southern Arabia was shipping more than three thousand tons of incense each year to the Mediterranean world. Even as they were trading by sea, Arabian traders sometimes stored their best aromatics inland in cities protected by hostile stretches of desert such as Petra, in modern-day Jordan, where they were less vulnerable to thieves.

Knowing the high value of their product, traders in frankincense guarded their secrets with legends about the mysterious trees. Said to be protected by large and fierce red snakes that would leap into the air to attack any intruders, the trees were believed to grow in an area of diseases and epidemics, making them dangerous to harvest. Or, in a different story, the mythical phoenix was said to nest in branches of the tree and feed on its tears. At an ancient oasis, one of the stops on the trade route, the Lost City of Ubar flourished during the incense trade. Referenced in the Koran and called by Lawrence of Arabia "the Atlantis of the sands," this city cast a spell that lasted centuries. Trade through Arabian cities and oases prospered until the Greeks discovered a way to bypass the long and dangerous land route through the peninsula to sail the Indian Ocean laden with incense. The character Sinbad the Sailor may have been based on merchants who sailed around the Arabian Peninsula and traded in frankincense. Sinbad's adventures appear in the collection of Middle Eastern folk tales *One Thousand and One Nights,* translated as *The Arabian Nights* by Sir Richard Burton in Victorian times.

Frankincense has also symbolized excess and luxury. The emperor Nero is said to have burned an entire year's harvest of frankincense on the death of his favorite concubine, or was it his wife? At the time of Christ's birth, the Roman Empire was importing some three thousand tons of frankincense each year from controlled supplies in the Middle East. When Alexander the Great was young, he once scattered handfuls of frankincense on the altar to burn as a fragrant offering to the gods. His tutor, Leonidas, chided him for wasting the precious aromatic and told Alexander that when he conquered the lands from which frankincense originated, he could afford to be extravagant with it. When Alexander conquered Gaza twenty years later, he discovered a large stash of incense and sent the now-elderly Leonidas a generous gift of frankincense and myrrh. Alexander was, however, confounded in his search for a source of frankincense and never made it through the harsh deserts between Gaza and the southern end of the Arabian Peninsula. The same could be said of Caesar Augustus, who sent ten thousand troops to southern Arabia in 25 BC to obtain the treasures of Arabia Felix, only to be defeated by the harsh environment. For perfume and incense, the best of Arabian frankincense stays within Oman for use by the royal house. The Arabian perfume house Amouage is located in Oman and uses frankincense and other traditional Middle Eastern perfume ingredients to create what it has called "the most expensive perfume in the world." Founded by an Omani prince, Amouage celebrates the ingredients particularly found in Oman.

Frankincense is mentioned many times in the Bible and is closely tied with myrrh—also a lovely incense ingredient. The gift of the Magi to the Christ Child of gold, frankincense, and myrrh is thought to symbolize: his kingship—gold; his spiritual nature—frankincense; and his death—myrrh. Orthodox and Roman Catholic churches of Europe and Latin America are some of the largest users of the Eritrean type of frankincense and use the formula of ten-fifteenths frankincense, four-fifteenths benzoin, and one-fifteenth storax for their incense. This blend is burned in a thurible, an elaborate censer that is swung ceremonially to release the smoke and bless a space or sacred object. A censer is a fireproof cup with a layer of sand or fine gravel for holding a specialized charcoal puck. Once the puck is red-hot, frankincense (or any resin or resin blend) is placed on top to offer up its sweet and smoky fragrance. Alternatively, the resin may simply be melted in a heater to achieve a less smoky, more pure frankincense experience.

As I live a life filled with aromatics, frankincense gives me joy and peace. There is something simple in the sacred resin that conveys the feel of a tree that has withstood the test of time and environment to create joyful and healing tears. For creating perfumes, I work with various extracts, mainly the steam-distilled essential oils that exhibit the same fragrances but without the immediacy and impact of the heated resin. Because much of the beauty of frankincense and myrrh comes out through heat, I love making beeswax candles using a combination of resin and essential oil. Working with frankincense reminds me that the word *perfume* comes from the Latin *per fumum,* meaning through smoke. Frankincense essential oils and extracts are valued for their complex aromas of resin and citrus and the way they enhance and lengthen the effects of woody and floral ingredients. Working with myrrh requires a lighter hand unless I am going with a full-on resin-y, incense-y blend. I love how myrrh can add sweetness and roundness to woody bases in perfumes. Despite a tendency toward the plastic and medicinal, a fine myrrh essential oil has a balsamic essence that is also warm, spicy, and strangely comforting.

The New World is also home to spiritual resins. In the Americas, copal plays similar roles to frankincense and myrrh in purposes both botanical and anthropological. The word *copal,* derived from the Aztec word *copalli,* refers to scented resins from a variety of trees and shrubs known as torchwoods. These trees in the genera *Bursera* and *Protium* range from the southwestern United States to Mexico and Central America and into South America. Most torchwoods used for copal are in the Burseraceae family—the same as frankincense and myrrh, although sometimes the fragrant resin of pines (*Pinus* spp.), legumes (*Hymenaea* spp.), or sumac (*Rhus* spp.) is used. The genus *Bursera* contains more than a hundred species that occur from the southwestern United States to Peru and can be dominant in seasonally dry tropical forests, desert and oak savannahs in southern Mexico, and some moist forests. Torchwoods range in size from small shrubs to tall trees, and many have brightly colored succulent trunks of blue, yellow, green, red, or purple with outer bark that flakes off in colored sheets. In Florida, the gumbo limbo tree (*B. simaruba*) is sometimes called the tourist tree because of its red and peeling bark—reminiscent of tourists who overdo their exposure to the tropical sun. Resin from *Bursera* trees forms in a system of canals in both trunk and leaves: it is rich in terpenes that contribute to the fresh, lemony, and resinous fragrance while protecting the plant. Flowers are pollinated by a variety of insects, and birds feed on their small, often fleshy fruits, including the white-eyed vireo (*Vireo griseus*), which establishes and defends territory around the trees in its wintering grounds. The Mexican linaloe tree, *B. linanoe,* is known for its floral fragrance and today is cultivated in India for its essential oil, which may be obtained from the fruit instead of wood. *Bursera bipinnata* is another source of copal and has a fresh piney resin. *Protium* species tend to favor the wetter forests of Brazil, where the incense tree (*P. heptaphyllum*) is found. *Protium copal* is one of the main sources of copal incense and is found in Mexico, Guatemala, and Brazil.[14]

Perhaps a more useful classification of copal resins is based on color. Copal blanco may come from exuded resins and is light in color; copal negro is dark and often comes from beating the bark of resinous trees. To some, dark copal was the blood of trees and seemed a fitting symbol for blood rituals, whereas light-colored copal resembled rain drops and was used to call down rain and scare away "bad winds." Both Mayas and Aztecs used copal as incense in their ceremonies but also to cleanse their living spaces, to ward off evil, and to bless weddings and births, farmlands and hunts. Copal was food for the gods much like maize was food for people: corn-cob shaped offerings of copal resin have been found as religious offerings. The use of copal incense in sacred ceremonies has persisted through the centuries despite early Christian dismissal of the native resin and importation of frankincense from overseas for use in churches.[15]

The sticky resins and terpene constituents of copal trees are both repellent and attractant to the insect communities of their ecosystem. There is a small weevil that chews into the bark of *Protium* species to release resin and use the resulting sticky blob as a nest for its larvae. Once the resin is released and weevils are using their share, a careful observer may find assassin bugs daubing a bit of resin on their legs to assist in the capture of stingless bees that harvest the resin for their nests. Ants and orchid bees also drop by to collect their portion for nest protection (ants) or perfume-making by orchid bees (more on this in chapter 8). Brushing aside—or picking out—the insects, *copalleros,* or copal collectors in Mexico, bring their special knives to scrape resin into the leaves of an agave plant.

But it is a group of small beetles called flea beetles in the genus *Blepharida* that dramatically demonstrate the defensive properties of these resinous plants—the aptly named "squirt-gun defense." When a small insect chews through a vein in the leaf of some *Bursera* species, resins will flow copiously or even squirt out, trapping the insect or gumming up its mouthparts as the resin solidifies. Some plant species maintain the resin under pressure in the leaves so that any rupture of veins in the leaves results in a squirt of sticky and aromatic resin targeted directly at a chewing insect. Larvae of the flea beetle, through a long coevolutionary relationship with copal plants, counteract this defense by draining the resin before it can squirt. They will bite through large midveins or cut trenches in the leaves to reduce or redirect the secretions. Whether they trench or bite depends on the leaf architecture. If smaller veins bifurcate off one main vein where biting through the large vein will drain the leaf and eliminate the squirt, then that is the behavior. Some veins are more netlike, and this is where the trench behavior is used to stop or redirect the flow from several veins. Insects that get squirted will abandon the leaf and remain inactive for several hours before moving on to another leaf. Or sometimes a younger, inexperienced larva will die covered with resin. Although the insect can bypass the leaves' defenses, it may experience increased handling time and decreased survivorship, effects that potentially reduce overall damage to the plant. Adult flea beetles feed on the resinous leaves and accumulate terpenes in their bodies, which they can use to their benefit. How, you ask? By creating fragrant feces to deposit on their backs and thereby forming a fecal shield against predators. Or, if attacked, they can release a spurt of repellent terpene-rich secretion either anally or orally.[16]

I have not been to the frankincense forests of Oman or seen myrrh trees in the Horn of Africa, but I have spent much time in the deserts of the southwestern United States. On one hike in the foothills of Utah mountains, I was introduced to olfactory botany by a fellow scientist who taught me to differentiate pines by scent. Since then, I often close my eyes to inhale the scent of the desert, and I have learned the fragrance of pinyon pine resin and tough little creosote bush leaves. These are not copal plants, and I hope someday to seek copal out specifically, but these plants are rich in terpenes, and I, like the little flea beetles, have been thoroughly and fragrantly gummed from time to time.

Last, a word about another famous resinous plant. Resin, fiber, and food come from cannabis, which may have been cultivated as early as 10,000–3000 BC in China. Very much a weed, cannabis can grow almost anywhere, including in a small room with grow lights, but it prefers moist areas along riverbeds. Before the plant was known for its psychoactive properties, its parts were used as anesthetic and incense in ancient China, its fiber for sails on wooden ships and for paper, and and its seeds to grow food in times of famine. There are several types of cannabis, variously referred to as *Cannabis sativa, C. sativa* ssp. *indica,* and *C. sativa* var. *ruderalis,* that are different in appearance and use. The *sativa* type, originally from Europe, is widespread, with a wide and gangling habit; *indica,* most often used for hashish, is shorter and denser and originated in Asia; *ruderalis,* likely the product of the combined gene pools of feral plants, is short and scrawny and a hardy pioneer. Because cannabis has been cultivated for so long, has a very flexible and adaptive genome, and likely traveled with a variety of people from earliest times, it is hard to pinpoint an origin, but research points to Old World northern temperate climates.[17]

Hollow stems with fine hairs produce strong hemp fibers useful for rope and tough cloth, seeds can be pressed to yield a burnable oil for lamps or making soap, and female flowers produce an abundant resin that may be burned for incense, smoked, or eaten to achieve its mind-bending effects. Plants invest in the resin and in constituents like cannabinoids and terpenes perhaps as a shield against hot and dry conditions, a way to capture pollen for the female plants, or protection against diseases and herbivores. As humans have used it, they have defined it as marijuana for the flowers and leaves, sinsemilla for the flowering tops, hashish for the resin, and hemp for the fiber. They have smoked leaves and flowers using a water pipe or blended with tobacco, included them in food and drink, and burned the resin as incense. The Chinese were early believers in the medicinal qualities of marijuana, as was Brownie Mary, who lived in San Francisco in the latter half of the twentieth century. Mary Jane Rathbun, volunteer and marijuana activist, made brownies with marijuana—first to sell and then as a volunteer for cancer and AIDS patients to ease their nausea.

Beginning with the Chinese, hemp was valued for making paper, a secret that was uncovered by Arab traders and brought back to the Middle East. By about AD 400, Saxons and Vikings in Britain were growing and using hemp extensively for a variety of purposes. The fiber brought riches to Venice, whose hemp workers made the finest rope in the world for three hundred years during the Serene Republic's days as an oceangoing power on the Mediterranean. Plants were grown in the American colonies and used for the paper on which the first draft of the Declaration of Independence was written, and prairie schooners crossed the Plains covered by cloth made of hemp. Cannabis has strolled along with gods and humans in legend and history, including Hindu deities—especially Shiva, who is often called the Lord of Bhang; 1960s hippies who followed in the steps of the Sufis on the Hashish Trail from Europe through the Middle East; Muslim assassins of legend; European writers such as Victor Hugo, Alexandre Dumas, Oscar Wilde, and W. B. Yeats in the nineteenth century; jazz musicians including Benny Goodman, Cab Calloway, and Louis Armstrong in the 1920s and 1930s; and U.S. soldiers in the Vietnam War. In the United States marijuana was grouped with other psychoactive drugs and criminalized until the twenty-first century as states began to legalize its medicinal and recreational use.[18]

Resin-producing structures on the vegetative and floral parts of cannabis plants are called capitate trichomes and are shaped a bit like a mushroom with a microscopic resin-producing ball atop a thick stalk. The ball contains the resin and releases it along the leaves, stems, or flowers in response to damage or as a protective coating against desiccation. To supplement the resin, cannabinoids and terpenes in the plant are thought to prevent damage from herbivores, but some pests can get through the defenses. Tiger moth caterpillars (*Arctia caja*) prefer to feed on the high-THC (tetrahydrocannabinol) variety of cannabis and can store the toxins in their body to repel predators. The plant's more than five hundred compounds include terpenes, cannabinoids, flavonoids, and others. Delta-THC is a cannabinoid, a type of nonvolatile chemical unique to the marijuana plant. Alone, THC is odor free but may comprise a large part of the sticky resin and is responsible for the psychoactive effect. Terpenes and sesquiterpenes famously give different varieties of cannabis their characteristic odors and tastes: they vary with geographic range and with breeding as growers specialize in new varieties and constituents. Although not psychoactive, different terpene flavors and fragrances seem to affect perception and preference. This interaction of constituents like cannabinoids and terpenes is often called the entourage effect and may be summarized as thinking of the plant being more than the sum of its parts.[19]

Agarwood (Aquilaria sinensis) *arils and wasp with flower*

2

Fragrant Woods:
Agarwood and Sandalwood

The Wood of the Gods is a product of the light-colored wood of trees called agarwood in the genera *Aquilaria* and *Gyrinops* and found in forests from China to New Guinea. They are not important timber species—the bark may be used to make paper or twine and sometimes boxes—but sometimes a tiny invading fungus, a small injury, or perhaps a boring insect sets in motion a mysterious protective process that produces a dark and aromatic resinous feathered pattern within the living wood. The result is a highly valued aromatic wood called agarwood.[1] Also known as *gaharu*, aloes, *jinkoh,* oud, or eaglewood, dense, resinous wood from the tree is the legendary sinking wood of ritual, incense, medicine, and perfume. Tiny chips of agarwood are burned for incense and ritual and may be distilled to produce a highly fragrant and distinctive essential oil.

Agarwood or oud essential oil is an ancient perfume and incense ingredient in the Middle East that has come more recently to European and American perfumers who wish to add depth, longevity, and mystery to their blends. In describing the scent of the essential oil, the perfume industry may use words like woody, intense, animalic, with notes of tobacco and leather. Sometimes, however, oud is not entirely proper, and the smell is more accurately described as eau de barnyard, with notes of well-used straw from a stable with nuances of musk and sweat overlaying charismatic wood and leather notes. Some varieties are more approachable and have the occasional berry, spice, amber, Band-Aid, or grassy aspect. Whichever note comes to the fore, it is the complicated and sensual nature of the fragrance that devotees find so enthralling. If the wood itself is burned or heated, it releases two types of molecules, sesquiterpenes and chromones, which together act to produce a fragrance different from either wood or essential oil. Sesquiterpenes welcome the heat, while chromones release their subtle scent in smoke.[2]

When resinous, agarwood is one of the most valued woods in the world, and the essential oil is one of the most expensive perfume ingredients. Not all agarwood trees in a forest or even all trees within a genus produce fragrance, perhaps only 10 percent, confounding those who would harvest the trees. The fragrance of the resin within the wood results from perhaps as many as 350 different constituents that sometimes accumulate to the extent that the resin is nearly black and an agarwood chip will sink when placed in water. A product of pathology, the defensive resin is a result of wounding paired with microbial invasion. Certain fungi called endophytes are strongly associated both with the formation of the resin and with the characteristic agarwood fragrance. Agarwood trees everywhere have declined in number through overharvesting, endangering them worldwide, but there are projects in various countries to create plantations for harvest and to conserve trees in their native habitat. Success of the plantations depends on the ability to produce resinous wood and on a shorter timeline than in the wild. Countries within the historical range of the species are experimenting with plantation-grown wood, most since the late 1990s or early 2000s, and include Bangladesh, Bhutan, Cambodia, India, Indonesia, Laos, Malaysia, Myanmar, Papua New Guinea, Thailand, and Vietnam. To induce the formation of scented resin, trees are wounded or injected with a fungus, often both. The hope is that such techniques will consistently produce resin-yielding trees more predictably and quickly than the decades often required in the wild. Plantings have focused on trees in the genus *Aquilaria: A. malaccensis* in India, *A. crassna* in Cambodia, Thailand, and Vietnam, and *A. sinensis* in China. Australia is venturing into the agarwood business with *A. crassna.* Flowers of agarwood trees may be white or yellow-green; they bloom infrequently, open at night, and have a sweet, floral fragrance that draws in a variety of insects, including their primary pollinators, moths. Seeds develop inside woody fruits covered in gray hairs that split when ripe, but the seeds within do not fall to the ground; instead, they hang like small green pendants from delicate hairs still attached to the capsule. A small, fatty body called an eliasome is an integral part of the seed, and it attracts hornets (*Vespa* spp.) that detach the seeds and carry them away to eat the nourishing fat. Agarwood trees often have poor seed dispersal on their own, and this relationship with hornets ensures that not all seeds drop to the ground directly under the mother tree but some get dispersed as much as 260 feet and more away.[3]

Agarwood trees grow naturally in forested areas from northern India through much of Indomalaya and are sparsely distributed. The few old trees that remain are mostly in preserves or virgin forests. In Borneo and parts of Malaysia agarwood trees can be found deep within mixed hardwood forests, often on steep ground near rivers that bisect the landscape, up to about three thousand feet in elevation. These challenging landscapes have protected the trees, in some cases, and provided income for local people. In Hong Kong, whose long relationship with agarwood and the incense trade is reflected in the city's name, which means fragrant harbor, there are natural populations of agarwood trees in the mountains and around the countryside as well as over twenty million planted and cultivated trees. Hong Kong's Agriculture, Fisheries, and Conservation Department protects young trees with fencing and plants seedlings in country parks. In home gardens of upper Assam in India, agarwood may contribute as much as 20 percent of a family's income, and trees are being grown for shade in tea plantations. Although few agarwood trees remain in the wild, some estimates list nine to ten million growing in plantations in northeast India alone. It is difficult to determine which trees have developed the resin, but collectors in Assam have traditionally classified different forms of the species *Aquilaria malaccensis* and understand the potential of each form to produce high-quality resin from infection. With long years of experience, they have signals that they look for, including obvious signs of disease like decayed branches, dieback of top and outer branches, ants populating fissures in the bark, and a yellow to brown tinge in the wood under the outer bark. But almost as telling are signs of a borer insect, *Zeuzera conferta,* that creates lesions and tunnels within trees and is associated with a higher than usual occurrence of the resin. Resin is collected between January and April to obtain the most material with the finest odor, and some collectors will chip out the diseased parts, leaving the tree intact.[4]

In central Borneo, the Penan Benalui people have been harvesting agarwood from mountainous hillsides for centuries, giving them an understanding of where to search within their forests. Such on-the-ground wisdom is sometimes called traditional ecological knowledge and is a subject of ethnobiological research aiming to understand local involvement in and knowledge of natural vegetation for human use. To harvest the fragrant heartwood of the agarwood or gaharu tree requires knowing how to find the fungal infection in the first place. Unlike Assamese growers, Penan men must enter deep forests where the signs are more subtle and where stealth rules. Before the hunt, they don black. They do not speak of where they are going because someone may be listening. They seek the fragrant, dark, nearly black resin of the *Aquilaria* tree hidden within tall trunks and under thin gray bark. Sleeping in small camps with black tents and equipment—

black is the color of gaharu—they search largely in silence, listening and looking for signs passed down as lore through generations. Because resin is in the heartwood, they must look for subtle physical signs to indicate its presence in less than 10 percent of trees: in this way they combine ritual with accumulated knowledge. Cash value from trading the resin has allowed Penan Benalui to purchase trade goods not available otherwise, and that makes it well worth the search. So they listen to cicadalike insects associated with the trees, examine the landscape for patterns, and hedge their bets by using items of black that mimic the dark resin. Some local groups will slash a tree or its roots to determine whether the valuable resin is present and then extract the resin without killing the tree, leaving it to continue growing and produce more resin.[5]

Penan Benalui understanding goes beyond lore and, it turns out, reinforces some of what scientists have learned. Indigenous groups recognize the clumped nature of *Aquilaria* distribution because seedlings tend to grow near a mother tree; they know to search the drier and higher small streams for resinous trees where trees may be stressed; they recognize palm trees that tend to grow in association with *Aquilaria;* and they confer upon the spirit of the cicada the origin of the scented resin. They also recognize the symptoms of disease, such as insect boreholes, leaf drop, growths, and a hollow sound when thumped. What does science say? It tells us that trees growing in more challenging environments may be more easily damaged and infected by the fungus and so produce the aromatic resin more quickly. Although the trees are widely distributed, they tend to be clumped due to low dispersal, despite the help of the wasps. This clumped pattern may create more likely opportunities for infection among the older trees. And because cicadas have a life cycle that goes from soil to tree and back to soil, they are excellent vectors for fungi that thrive in the forest floor.

Resinous agarwood has been used as a medicinal ingredient for centuries, including by the Greek physician and botanist Dioscorides writing in 65 BC, and as a perfume ingredient called aloes, as mentioned in the Psalms and the Song of Solomon in the Bible. The complexity and emotional resonance of the scent of incense means that it has long found a place in ritual, as described by Sanskrit texts dating from 1400 BC. Buddhist monks have for centuries burned the resin during meditation and prayer; when agar resin is used in prayer beads, the warmth of fingers releases its fragrance. The Arab world, despite not having native agarwood trees, has been an important consumer of the resinous wood and its products for over two thousand years, and it was likely an item of trade along both the Silk and the Incense Roads. Like sandalwood and scented resins, agarwood was a portable product that retained its scent—and value—over weeks and months of travel between source and destination. The Prophet Mohammed appreciated the scent of agarwood in perfumes and for scenting his clothing but also as medicine. The tales of *One Thousand and One Nights* include references to scenting houses and palaces with the fragrance. Middle Easterners do not just use a spritz of perfume but will scent clothing and homes with incense or sprays and reserve the best quality for special occasions and weddings.

In the Far East, incense was originally associated with Buddhism, where it was thought to assist in invoking the Buddha's presence to bring a more peaceful world. This incense was a blend of up to seven ingredients, including jinkoh (agarwood), sandalwood, cloves, cinnamon, and camphor, that was heated to release the scent. In China beautiful censers and incense-related products were created for the art of appreciating the scent of agarwood, and in Japan the fragrance may have wafted from the flowing sleeves of silken robes or scented a quiet room set aside for meditation or ceremonies. Chipped resin was blended with other fragrances such as ambergris, cloves, and sandalwood for a kneaded type of incense to burn in winter, while the chips themselves would be burned or heated in the summer.

In perhaps the world's first novel, *The Tale of Genji* by Murasaki Shikibu, written in eleventh-century Heian Japan, incense is a vehicle of pleasure as well as religion and a central part of the life of an aristocrat. People of refinement were known by their crafting of incense blends and would count their skill at blending on a level with other arts such as music or poetry. As will happen, art evolves and incense practitioners moved to burning wood, specifically jinkoh, to appreciate and identify. Eventually there came the incense ceremony called Koh-do (you may also see it spelled as Koh-doh or Kōdō), or listening to incense. Koh-do invites participants to experience or listen to incense as they follow the guidance of a composer who selects perhaps three types of incense for listening, after which participants match an unknown sample to one of the three in the basic form of the ceremony. In some versions, descriptors of the fragrance ascribe personalities to each type that may derive from historic locations but may also involve terms associated with taste. For example, the best is Kyara, with a dignified smell that is also slightly bitter, reminding one of an elegant aristocrat. Contrast that description with Sumotara as possessing a sour beginning and end that may be mistaken for Kyara until one recognizes its ill-bred nature, like a servant passing as a noble. The phrase "listening to incense" was first used in the fifteenth century, perhaps derived from Chinese, and made uniquely Japanese. As a ceremony, it incorporates dedicated paraphernalia, a Koh-do master, rules, and records. As a game it is played with others and is intended to be an enjoyable social interaction that may be as simple as identifying one of three incense types as described above or more involved with a story, a journey, or poetry as the theme for choosing and naming incense fragrances. The popularity of Koh-do peaked in the Edo period and began to decline in the nineteenth century as Japan opened to the West. But by the twentieth century, Koh-do masters began offering classes again, incense shops opened, and a wider audience was found, including among the young. Although Koh-do is highly ceremonial and has a secretive set of rules, participants attend with the anticipation of fun, an escape into the world of scent, and social interaction. Perhaps at the end of the day the answers matter less than the joy of sharing a fragrant experience with others.[6]

There are two legendary pieces of agarwood. The first washed ashore on the island of Awaji, Japan, in about the year 585. Upon burning the wood (or at least a portion of it), locals appreciated the fine smell of the smoke and presented it to the imperial court. The second was a gift from the Chinese to the Japanese emperor Shōmu. Called the Ranjatai, this legendary piece of resinous wood has been preserved at the Shōsōin repository for precious items in Nara and is taken out periodically for exhibition in rotation with other treasures. Small slices are removed, very rarely, as a special tribute, and the block of wood itself, weighing about twenty-four pounds and nearly five feet long, has small markers for each tribute piece. It is said to have the perfect fragrance.

Mysterious in origin, difficult to find in the wild, rare, exotic, and hugely expensive, agarwood has earned the name Black Gold. The absolute best of the wood—the dark and deeply resinous sinking type—is saved for use as an incense, where small chips of wood are burned or heated. Lower-quality yet still aromatic pieces of wood are distilled for essential oil in a traditional process that involves long periods of soaking the wood followed by hours or even days of distillation. The soaking process releases the fragrant resin from pockets within the wood, and distillation allows aromatic molecules to rise with the steam to be gathered. Both wood and essential oil are evaluated and graded within the trade by experts familiar with a particular product, such as chips of resinous wood for incense, blocks of wood,

essential oil, accessories, or various value-added products like perfumes and carvings. Appearance and fragrance seem consistently to
serve as hallmarks of quality, but there is also the somewhat simple act
of placing the wood in water and seeing if it will sink because woods
thick with resin will be heavier than water, giving resinous agarwood
the epithet "sinking wood." Unlike many perfume ingredients and
essential oils that are regulated by industry, standards of quality for
agarwood may not be completely objective. Traditional grading systems, even in use today, appear to be mostly dependent on physical
characteristics like color, resin content, and weight of the resinous
wood. Country of origin and tradition may also influence evaluation,
and suppliers and traders in each country seem to have their own
standards for a product. For the essential oil, aroma and longevity on
the skin are both important. Fragrance chemists have begun elucidat-
ing the constituents that make up quality agarwood oil using GC/
MS—short for gas chromatography/mass spectrometry—a process to
separate and identify constituent molecules in a blend using tiny
amounts of essential oil. To understand the subtle difference in resinous wood for incense a GC/MS instrument can be used, but it is injected with smoke instead of liquid. This method has allowed
scientists to describe not only the complex array of sesquiterpenes and
other fragrant molecules found in the wood but also those compounds called chromones that bloom when heated and, with sesquiterpenes, create a rich, sweet, warm, and long-lasting incense aroma.[7]

In the forests of southern India a tree grows old and its heartwood
becomes a rich brownish-red and deeply fragrant.[8] Sandalwood trees
(*Santalum* spp.) produce a protective and aromatic oil deep within the
tree that grows more scented and abundant with age. As a tree grows,
active sapwood is produced in a layer on the outside of the trunk,
roots, and branches under the bark where it stores nutrients and water

to send from roots to leaves as well as responds to injury by growing tissue or producing defensive compounds. Inner layers of sapwood eventually die in a natural progression as the tree grows and are converted to inner wood, aptly named heartwood, that serves to support the tree. In many trees including sandalwood, the heartwood becomes darker and more fragrant as volatile chemicals are produced at the boundary of sapwood and heartwood and work their way inward. In sandalwood, an essential oil is produced that is one of the most valued in the world.

Age produces beauty deep within sandalwood trees as scented molecules are produced that concentrate in the oldest branches, trunk, and roots. The fragrant molecules, mainly sesquiterpenoids, are the trees' defense against pathogens and chewing insects. While young, sandalwood trees need botanical companions: the sprouted seedlings grow best when connected with the roots of nearby plants, sharing nourishment, thus earning for themselves the nickname "vampire trees." Yet the sandalwood tree also taught the famed Bengali poet Rabindranath Tagore a lesson in love when he wrote, "As if to prove that love would conquer hate, the sandalwood perfumes the very axe that lays it low." Rather than a vampire that sucks out life, the tree gives back and produces aromatics that are supportive, healing, gentle, and yet strong. Sandalwood is fragrance made visible when used for beautiful traditional carvings by the Gudigars, traditional artisans from the Uttara Kannada district in Karnataka, India, or used as a fragrant paste to adorn Hindu gods. It is venerated and used in three religions: Hinduism, Buddhism, and Islam.

High-quality sandalwood is ground into fine pieces to be distilled, often using techniques and tools that have remained the same for centuries. The scent of both wood and essential oil is deceptively soft, with buttery notes, skin tones, and elegant woody aspects, yet it lasts and lasts, making it an important base note in fragrance compositions. The accepting nature of sandalwood oil also makes it the perfect liquid vehicle for creating attars: the product of delicate florals distilled directly into sandalwood essential oil. In my collection are a variety of sandalwood essential oils from different species, including a few precious ounces of endangered *Santalum album* from India, usually called Mysore sandalwood, that serve as a standard for comparison with other types and species. Sandalwood asks you to be patient. At the beginning it may appear that there is nothing on the scent strip, but a moment of quiet will reward the nose with a fragrance that is so rich and complex that you may not be able to come up with terms to describe it. To my nose the odor is very slightly balsamic (meaning a sort of resinous caramel) and smooth with precious wood overtones: the scent has a charismatic presence that is hard to describe. A few more moments and you begin to appreciate its elegance, its depth and complexity, and how it may blend seamlessly with the animalic fragrance of skin.

Although Indian sandalwood is the most famous variety, Australia is thought to be the origin of the genus, with several long-distance dispersals.[9] Sandalwood species, about fifteen all told, occur in a swath of tropical lands from Indonesia in the East to, once upon a time, the Juan Fernández Islands off the coast of Chile, where the single native species is extinct, from Hawaiʻi to New Zealand, and to the Bonin Islands south of Japan, where a tiny population exists. *Santalum spicatum* from Australia has a bright and resinous woody fragrance that finishes with a light yet still elegant sandalwood note. Australia has built an industry around this species of sandalwood, but over the past couple of decades the continent has also supported new plantations of *S. album*. New Caledonia, Vanuatu, Fiji, and Tonga produce oil from *S. austrocaledonicum* that has a beautiful fragrance with, occasionally, a bit of vanilla sweetness, leading to a classic buttery, woody drydown (the final lingering element of a fragrance).

Vanuatu natives recognize two varieties of this species, differentiating between a "woman" variety that produces more heartwood early on, is shorter and fatter with more rounded leaves, and has many fruits, and the "man" variety that takes a while to produce good heartwood, is taller, has more pointed leaves, and bears few fruits. Fiji was an early source of sandalwood for Tongans, who would trade sting-ray spines, bark cloth, and whale's teeth for the wood. *Santalum yasi* also grows in these islands. Hawaiian sandalwood is called *'iliahi* by the locals and comes in several species, including *S. paniculatum, S. freycinetianum, S. ellipticum,* and *S. haleakalae.* I have a small bottle of *S. paniculatum* and find that it departs a bit from the traditional Indian *album* type. There is a hint of floral and a mysterious "old library" note that transitions to a lovely cedary-woody note before the nearly oudlike sandalwood finish. Once thought extinct, Hawaiian species have persisted in remote areas and higher elevations where collectors and traders in the nineteenth century did not go and where the trees have refuge from grazing animals.

Sandalwood's roots in India are a mystery—some think that the tree is not native but was introduced from populations in Indonesia over two thousand years ago. Perhaps early traders to Timor recognized the beauty and potential value of the species and brought back seeds or seedlings to India. Alternatively, the tree may have arrived from its native Australia in the gut of birds accustomed to flying long distances, such as the Pacific golden-plover (*Pluvialis fulva*) or Pacific imperial-pigeon (*Ducula pacifica*).[10] Despite its wide occurrence, sandalwood species throughout their range have been overharvested, and many are in danger of becoming extinct like the endemic *S. fernandezianum* from the Juan Fernández Islands. East African sandalwood, or *Osyris lanceolata,* is a fragrant sandalwood that may be used for essential oil and pharmaceuticals. Red sandalwood, or *Pterocarpus indicus* (synonym *P. santalinus*), from Southeast Asia may be used for cosmetics and in Ayurvedic medicine. Its lovely red wood is one of the preferred woods for the shamisen, a Japanese stringed instrument. The fragrant tree *Amyris balsamifera* grows in Haiti and the Dominican Republic: it is also called West Indian sandalwood or candlewood due to the high essential oil content that allows it to burn freely.

The finest sandalwood, sometimes called Mysore, is said to grow in southern India, in the provinces of Karnataka, Tamil Nadu, and Kerala, where it is currently protected from harvest through government control. Long history and close involvement in rituals, religion, and commerce have made sandalwood an important part of Indian culture and heritage. It was declared a royal tree by the sultan of Mysore in 1792 and in Tamil Nadu by the Madras Act in 1882. Indian sandalwood is listed as Vulnerable in the International Union for Conservation of Nature's Red List of Threatened Species. In addition to overexploitation of the trees for many years, spike disease, invasive weeds, fires, and grazing have brought the population to a small proportion of what once grew. In spite of its historical and cultural importance and value to local industries, including a large essential oil business in Kannauj, Uttar Pradesh, and sandalwood carvers in Karnataka, mismanagement by government agencies tasked with protecting the trees has resulted in overharvest, proliferation of illegal harvesting, and large-scale loss of trees.[11]

Despite laws protecting the trees and efforts on the part of various governments to ensure the safety of India's valuable sandalwood resource, banditry and thievery exist. The most famous sandalwood bandit in modern times was a man named Koose Munisamy Veerappan who spent his long career in the forests of Tamil Nadu, Kerala, and Karnataka. He is most noted for smuggling sandalwood and ivory and was viewed as somewhat of a Robin Hood by locals, although he did not hesitate to use intimidation and murder to carry out his thievery and evade capture. Veerappan was killed by members of a special task force in 2004 in an operation called Operation Cocoon. This was not the end of illegal sandalwood harvest and smuggling: a recent search of Indian newspapers turned up several accounts of disappearing trees. Thieves may appear in the middle of the night and brazenly cut down trees from places like the campus of Bangalore University, where officials have begun geotagging the trees and closing illegal roads. Two large trees were cut down and taken from local government quarters under the noses of security, who found only stumps, a few branches, and some foliage the next morning. Guards protect a forest in Kerala called Chinnar, which has a large natural stand of sandalwood, and there are plans to insert microchips into the trees.

Sandalwood trees, though not really vampire trees, are hemiparasitic, meaning that they can grow on their own but do not thrive that way. In a natural situation as well as in plantations, they do best if they can grow with a host plant that provides nutrients through a root connection. Young sandalwood trees produce a structure in their roots called a haustorium that will reach out and attach to the roots of nearby plants. Although a variety of plants can act as hosts, members of the Acacia family and other legumes seem to support young sandalwood trees best, providing an intimate connection between host and hemiparasite mediated through the root connections and extending throughout the plant community. As part of the local community, sandalwood flowers provide forage for local bees that are attracted to their sweet scent. In Australia the small, extremely rare marsupial the woylie (*Bettongia penicillata*), is known to cache seeds of the quandong tree (*Santalum acuminatum*) and native sandalwood (*S. spicatum*) in such a way as to encourage germination. In India, the Indian gray hornbill (*Ocyceros birostris*) eats the seeds and defecates them out to produce sandalwood seedlings in its nest middens. Birds overall seem to be important dispersers of the seeds, often dropping them in the middle of thorny brush, ideal for seedlings to reach out and find a nearby host.[12]

Sandalwood has a deep history of trade going back to about the third century AD and extending from Indonesia and India into China as well as in the Pacific Islands. Captain James Cook in 1778 was the first European to land on Hawaiian shores, and it did not take long before outsiders began loading the hulls of their ships with the fragrant wood. Beginning in about 1811, Western traders, including Americans, began trading with King Kamehameha I to acquire large amounts of the wood for shipping to Canton. Traders from the United States shipped fine furs and ginseng to China by way of Canton in exchange for goods such as tea and silk and would stop over in Hawai'i, where they often picked up a load of local sandalwood to add to their cargo. As traders realized the value of their sandalwood cargo in China, they increased their demand for the tree, but Kamehameha I restricted harvest and kept the right to trade in the wood to himself, thus controlling the exploitation of the trees. He may have learned from an earlier famine, perhaps related to laborers harvesting sandalwood instead of growing food, to keep manpower available for work in the fields as well as for harvesting trees in the mountains. He also learned, after he was tricked as to the size of a shipload of sandalwood (thinking in terms of square dimensions rather than the curved hull of a ship), to have holes dug in the ground the size and shape of a ship's hull for determining the amount of sandalwood actually required for one shipload. These holes can still be seen in places around Hawai'i.[13]

After Kamehameha I died in 1819, logging was suspended for a time and then recommenced under his son Liholiho and local chiefs. Liholiho abolished some of the controls over trade and allowed various chiefs a share in the trees. Continued trade by Americans with Canton led to high demands for sandalwood and increased debts on the part of Hawaiian chiefs, who took out credit for trade goods in advance of providing sandalwood to American ships. Local chiefs continued to amass debt until the rule of Kamehameha III, when the first written law of Hawai'i was passed to require Hawaiians to pay off their sandalwood-related debt. Each man was required to deliver a half picul of sandalwood, which was about sixty-six pounds, and women had to hand over one handmade mat of tapa, or bark cloth. Because wood was harvested in the mountains and transported on the backs of men over rough trails to the harbors, men involved in the work would often develop calluses on their backs and were referred to as callus backs. This is a short version of the long and sometimes tragic story of sandalwood in the Hawaiian Islands. Although Hawaiian sandalwood has mostly disappeared from hills and mountains near shipping ports on the coast, trees can be seen on the 'Iliahi Trail in Hawai'i Volcanoes National Park, and trees for essential oil production are being grown on plantations in the state.

Australia has a long history of growing the native sandalwood, *Santalum spicatum,* in plantations for essential oil production, and growers are taking what they have learned to establish more recent stands of Mysore sandalwood, *S. album.* Sandalwood was recognized as a valuable commodity soon after the founding of Sydney as merchants began trading sandalwood with China for tea, about which they were, and are, passionate. The gold boom of 1880 to 1918 had men going to the northern part of the country, where sandalwood was a good alternative money-maker if gold became scarce. A sandalwood distillery was established just outside of Perth, in western Australia, and soon the oil was being exported to England, where it was put into capsules to cure venereal disease, a demand that increased with the beginning of World War I. It was also used as an antiseptic, as a perfume fixative, and in soaps. In 1929 the government enacted the Sandalwood Control Act, and a year later four companies got together to form the Australian Sandalwood Company. Today there are a number of companies growing and distilling both *S. spicatum* and *S. album*.[14]

Hawai'i, after a checkered past of trade in native sandalwood, is encouraging planting of the native trees in both home gardens and mixed-tree forests, where they can be intercropped with suitable host species. Another Pacific Ocean country, Vanuatu, halted export of sandalwood between 1987 and 1992 to gain control of the trade, and now local farmers can earn a direct income through sales of planted trees to licensed traders. Characteristics of sandalwood make it suitable for smaller gardens since it is of high value and somewhat small in size, even though it requires many years to mature. Women and children can be involved in production and add to cash income. India has begun experimenting with plantations, including in Karnataka, where the State Handicrafts Development Corporation, responsible for carved sandalwood goods, encourages growing the trees instead of tobacco to improve the lives of farmers and preserve the iconic tree. There seems to be slow progress in moving away from government control of natural-growth sandalwood into provisions for private plantations through supplying information, seeds, and seedlings for landowners.

In India, the traditional method of distilling sandalwood uses wood fires in an hours-long process with low heat over an open fire and bamboo, clay, copper, and leather equipment. This is said to yield a finer oil than the faster steam distillation, where the wood is distilled under pressure. Highest-quality oil comes from old heartwood, and the best is from roots and the base of the trunk. Trees are cut down, uprooted, and sorted, then the heartwood is carefully separated and cut into chunky billets for sale and larger pieces for carving or, in the case of essential oil production, further ground into powder, fine but not so fine that it turns into paste in the still. In the traditional process the precious wood is placed into a copper *deg,* or distillation unit, that is filled with the appropriate amount of water and placed over a care-fully tended wood fire. The deg is connected by way of a bamboo pipe condenser (*chonga*) to the clay receiver, or *bhapka,* that is held in a trough of cool water. Once distillation is complete, the mixture is poured into a leather bottle called a *kuppi* to allow the mixture to settle and any remaining water to evaporate through the leather. Tra-ditional distillation is also used to create attars, where the gentle and supportive nature of sandalwood is used to hold the fragrance of a variety of local flowers. Distilled in Kannauj, their names sound ex-otic to Western ears—they are attars of *gulab,* or rose; *motia, chamelia,* and *juhi* describe jasmine varieties; and *kewda, champa,* and *genda* refer to particularly Middle Eastern botanicals such as pandanus, champak, and marigolds. In chapter 6 I describe *mitti* attar—the scent of rain distilled into sandalwood from the earth in Kannauj.[15]

Sandalwood appeals to all our senses, not just as a fragrance for perfumery and incense. A light-colored paste made from the pow-dered wood is used in cosmetics, and powder may be applied in visi-ble stripes to the forehead of followers of Krishna. Pacific Islanders traditionally used it to scent coconut oil that was rubbed into tapa or as a perfumed lotion for hair and skin. Statues of the Hindu god Ven-kateswara sometimes have a white forehead of camphor with a con-trasting streak of a musk and sandalwood blend. Beautifully grained wood seems made for carving into intricate objects but is also made into smooth beads that are pleasant and calming to the touch. The taste of sandalwood is bitter, its effect cooling. In Hawai'i, the wood was used to make the simple stringed mouth bow called the *'ukeke.*

What makes sandalwood essential oil so precious? We can find such descriptive words as deep, rich, precious woods, buttery, skin-like, and supportive or lasting, but perhaps, as with beautiful art or a complex landscape, it is best to accept that it is what it is. What it is, is sandalwood. Its fragrance does not come from injury, nor does it come from disease; its constituents are not aggressively defensive but simply a product of age and from the tree's being in the place where it belongs. There are technical analyses, and as with most essential oils, there are many constituents, but there are two alcohols, α-santalol and β-santalol, derived from the sesquiterpene santalene, that are standards that the industry uses to determine quality. Beta-santalol also has an interesting quality that professional perfumers have used from time to time to evaluate a perfume formula. Beta-ionone is a perfume constituent that imparts a beautiful violet aspect to perfumes but also has a strong woody note, especially when undiluted. A perfumer trying to determine the presence of β-ionone in a perfume can use β-santalol, which has a strong woody odor, to numb the nose to the fragrance of wood. Once she or he no longer smells "wood" from the β-ionone, the more ethereal violet notes come through, as they would in a perfume that evolves on the skin.[16]

Ancient or modern, Eastern or Western, there are a few things that remain true about incense. It is fragrance made visible, it rises to the heavens to carry messages to the gods, its pleasing fragrance connotes goodness and purity, and it pretty much always involves sesquiterpenes. Incense is still in use today throughout the world, in religion as frankincense and copal, or in intimate spaces like home. I like to envision a scene that might be from today or from four thousand years ago. The scene begins with a woman in a space, an intimate and personal place where she can meditate or pray, a place where the outside world does not belong. Early on any given morning, she gets out a precious abalone shell that she has used many times before and shapes sand in the bowl of the shell into a small mound. From a small, cloth-wrapped package she takes out two or three little pieces of a watery green frankincense resin and from another package a bit of flat charcoal. Grasping the charcoal with small tongs, she holds it in a flame for a few seconds, blows until the charcoal is burning uniformly and covered in a white ash, then places it atop the pile of sand. Then carefully she adds her bits of frankincense to the hot charcoal. Fragrant smoke immediately rises and with it the scent of resin, citrus, wood, and sweet balsam. Today she does not feel the need to do anything other than sit in this quiet space, breathe in the aromatic smoke, and center her thoughts. Another day she may use the smoke to purify and cleanse her living space. Hers is a timeless ritual as ancient as fire and as immense as religion.

SPICES

Small enough to fit into little bottles on a kitchen shelf and yet huge with fragrance and history, spices have had a worldwide influence on trade and exploration. At one time their lands of origin were the subject of secrecy and legends kept by the traders who helped establish empires and created vast wealth. We may think of spices as seeds, but they are fruit, sex organ, bark, leaves, and, yes, seed. Each produces a characteristic fragrance and taste from a suite of molecules that are often antimicrobial and protective in nature. We cannot always describe the fragrance of a particular spice, but nearly all of us will recognize the sharpness of freshly ground black pepper, the comfortable scent of nutmeg, or the zest of ginger and would find it hard to recognize a world without the complexity and interest they add to our foods (and perfumes). Many cultures have considered spices to have medicinal properties. Europeans especially found spices to be a revelation in adding dimension to formerly bland foods, prestige to grand feasts, and, possibly, sensual delight to the bedroom as purported aphrodisiacs. Each spice in our story has a different origin and biology, but all share a place in the history of the world. Sometimes we get a hint of their tales just by reading the exotic place-names on labels—I know I have asked myself about the difference between Tellicherry and Malabar when perusing bottles of black peppercorns at the store. For many centuries, traders traveled thousands of miles by caravan and ship to trade in the exotic and elusive wealth found in tiny portable form. Black pepper, nutmeg, ginger, cinnamon, and cardamom—spices filled the hulls of trading ships that sailed monsoon seas and rode the backs of camels across arid lands to deliver the precious cargo. A cargo that built empires.

The rise of Islam in the seventh century coincided with the growing power of the Middle East as religion followed the trade and communication routes. By the eighth century the Muslim world spanned the Himalayas to the Atlantic and trade routes, oases, and ports along the way. Muslims unified the Silk Road and built their success on the ability to keep sources a secret and on the tall tales they told of danger: crocodiles that lived in pepper swamps and had to be driven away with fire, giant eagles with nests of cinnamon, and the phoenix that nested in frankincense trees. In time, power expanded from Muslim traders to the city-states of Italy, premier among them Genoa, Pisa, and Venice in the eleventh and twelfth centuries. This was also the time of the Crusades, and Venice was right there to provide goods to the crusaders since it was ideally situated between East and West for trade along the Mediterranean. Spices began to arrive in greater quantities to Venice and hence to broader Europe, bringing huge profits with them, but Arab traders still dominated the trade routes. Active and booming ports for international trade sprung up in Sumatra, the Malay Peninsula, and especially the Malabar Coast of southern India.

Disease, too, moved along trade routes. In 1346 the Black Death, one of the deadliest plagues in history, spread rapidly through Eurasia. When the devastating pandemic ebbed in the early 1350s, populations had been greatly diminished, workers were in high demand, and wealth was somewhat more evenly distributed. Venice was then able to dominate the spice trade from Alexandria, Egypt, passing almost five million pounds of spices through its ports and on to Europe. Along with spices, exotic pigments used in paintings were arriving and supplied artists during the golden age of European art.

On the other side of the world spices, sandalwood, and incense continued to flow into China by the thousands of pounds. The origin of spices remained elusive to Europeans, providing motivation for the aspiring empires of Spain and Portugal to support sea exploration by such mariners as Christopher Columbus, Ferdinand Magellan, and Vasco da Gama, who promised to find the land of spices.

Although Columbus did not find spices, the Spanish conquistadors Hernán Cortés and Francisco Pizarro soon captured immense riches in the form of Aztec and Incan gold and silver, providing Spain with the wealth to become a global power. Vasco da Gama, sponsored by Portugal, followed a few years later with a carefully planned route to Asia around the coast of Africa to Calicut (now Kolkata), India, and he brought back actual spices. In 1519, Portuguese explorer Ferdinand Magellan set sail for the Spice Islands with a fleet of five ships under a Spanish flag, and although he did not live to complete the voyage, one of his captains, Juan Sebastián Elcano, circumnavigated the globe.

Dutch maritime traders followed Venice, Spain, and Portugal into the Spice Islands, and by the early 1600s they had created the Dutch East India Company. They soon expelled the Portuguese from the Spice Islands and moved on to control trade throughout most of the region. Dutch arts flourished during this time, with riches from the spice trade supporting artists including Rembrandt, Vermeer, and Frans Hals, as well as Delft artisans who created desirable blue and white ceramics. You may notice that this trade and wealth also represented a shift north, away from the Mediterranean countries of Spain, Portugal, and Italy and into northern Europe. Britain was not far behind the Dutch in building ships and sailing to the spice-producing lands under the newly chartered East India Company. In the centuries that followed, trade centers moved yet again as Russia expanded its reach into Asian trade. In a full circle, the discovery of oil in Persia led to another type of road—the one for trade in Black Gold—which brought new riches to the Arabian Peninsula, where it all began.[1]

For locals, spices like black pepper were medicine as well as cooking ingredients and a familiar part of the landscape where they are still grown in home gardens and nearby woods. For the wealthy and powerful they were also medicine as well as luxury. The Hellenistic king Mithridates the Great ruled northern Anatolia from 120 BC to 63 BC and was one of Rome's most formidable enemies. To survive the challenges of life and rule, he toughened himself as much as possible by taking small doses of poisons every day to develop a tolerance but also invented what he thought of as a universal antidote against poisoning called Mithridaticum.[2] Leaning heavily on spices and herbs, this medicine included, among other things, myrrh, frankincense, saffron, ginger, cinnamon, spikenard, balsam, lavender, long pepper, white pepper, carrot seeds, cardamom, fennel, and other herbs and plants, according to the 1746 *London Pharmacopoeia* (skink bellies were recommended in one version). This mixture was then blended in honey and wine. Later, Nero's physicians "improved" on the formula and included vipers as an important ingredient.

In my family we have a lot of memories and treasured family items from a year living in Sweden. Although I was young and have virtually no recollection of our time there, reminders of it filled our home with little painted Dala horses, warm wool gloves for skiing, and handmade woolen rya rugs. My mother had a recipe from our time there that she made for as long as I can remember—a cardamom-rich almond pastry that she would spend hours kneading and folding by hand for special occasions and at Christmas. Made with butter, almonds, and cardamom, the pastry filled the house with a wonderful fragrance that I have always associated with Scandinavia. The mystery of why this Middle Eastern spice has ties with northern Europe was cleared up for me when I read that the Vikings developed a particular fondness for the spice and brought it back from the East to be used in baking, in pomanders, and as an aphrodisiac. To this day cardamom means, for me, sweet and buttery pastries rather than curries and masalas.

These same spices that drove exploration and world economies are at home in the forests, mountainsides, and gardens of the Indian subcontinent, the Americas, and southern Asia. Black pepper vines dot the landscape of western India along the west coast where ginger and cardamom also grow under verdant vegetation watered by monsoon rains. Tropical nutmeg trees originate in Indonesia's volcanic Banda Islands, surrounded by coral reefs, where they thrive in the salty rains and sea winds. Cloves come from the nearby islands of Ternate and Tidore in the sea between the Philippines and Australia. From these little islands, shipload after shipload of aromatic cargo sailed around the world and to tables of the royal and wealthy. Saffron, one of the few nontropical spices, originated in the rocky slopes of the Himalayan mountains and around the Mediterranean, where the distinctive crocus flowers were gathered as early as the Bronze Age and memorialized in frescoes. Far from the Spice Islands, Mesoamerica's chocolate trees grew in forests, providing food for gods and rulers; vanilla, chocolate's boon companion, clambered up trees in the tropical habitats of the Americas.

Now is probably a good time for some terminology, beginning with the flower. Petals may be large or small, and they may be laid out in a lovely symmetrical arrangement called radial symmetry, like a buttercup or lily, or they may have bilateral symmetry, like a violet or an orchid. To test for radial versus bilateral symmetry, imagine drawing lines through the flower's face. If you can draw a line anywhere in the flower and have two identical halves, it is radially symmetrical. Think pies. If you can only draw one line to create a mirror image, the flower is bilaterally symmetrical like the human face. Petals form a corolla of, often, bright color that mainly serves to attract pollinators. Sepals grow below the petals to provide protection for the more fragile petals and are usually inconspicuous and green, but occasionally a flower requires a bit more show than petals alone, and then sepals come into play to assist in a bright and colorful display. When the petals and sepals look similar, they are referred to as tepals, and together they are called the perianth. Under the colorful show, flowers are the sex organs of plants. The male organ is called the stamen, which I remember because it includes the word *men*. It consists of an anther that produces the pollen and is usually held on a stalk. The female organ is the pistil, also usually on a stalk, and has a sticky end to hold onto pollen grains that will tunnel down through the supporting style to reach the ovary below and fertilize the ovules that become seeds. Seeds may develop within a fleshy fruit or may be mostly bare, but often they have some source of nutrition to help them get started. Seeds range in size from tiny like those in the bean of a vanilla orchid to the large coconut or somewhat less large avocado. Leaves are generally less complex than petals and are almost always green due to the chlorophyll that allows them to capture energy from the sun and convert it into sugars and starches necessary to maintain the plant. Given the abundance and diversity of plants and flowers, this is a simplified version, but it should get you started. For some plants, the synthesis of sugars, starches, and proteins as they take energy from the sun is as complicated as it gets, but the ones that go further, that manufacture what are called plant secondary compounds, are what these stories are about. While we generally think of flowers as the little factories that create fragrant chemicals, these same compounds are the constituents of taste in the seeds of a spice or the beans of a vanilla orchid. These are the compounds we have used for millennia as medicine, food, and fragrance and sometimes to alter our mood.

Cardamom (Elettaria cardamomum) *flower and leaves*

3

Spices of the Western Ghats

Rich, green, and mysterious, the Western Ghats range on the Malabar Coast of India were and are a significant source of spices and, one might say, the foundation of the spice trade. To the rest of the world these faraway spice forests once seemed like a magical place: legends told of snakes, bats, and giant eagles that protected the spices. Wet monsoon forests grow up to about four thousand feet above sea level along the flanks of the mountains and support a rich and diverse flora and fauna where structure is defined by layers. Tall emergent trees reach for the sun and provide filtered light for a secondary layer of vegetation. Nearer the ground are shade-loving shrubs, grasses, herbs, and climbing vines that fill in the spaces and use the forest structure to reach upward. Rainfall patterns are drivers of diversity in spice-yielding rain forests with between two and three hundred inches of rain, most falling between the summer monsoon months of June through September, followed by the dry season of autumn and winter.[1] These are the same wind-creating monsoons that traders once used to push their sails across the Indian Ocean to gather the bounty of spices and return home to sell. Three of the most popular spices in the spice trade originated in these rich and protected niches: black pepper, ginger, and cardamom. Vining peppers (*Piper nigrum*) grow up the trunks of trees and produce tiny white flowers followed by small green fruits in elongated clusters. Ginger (*Zingiber officinale*) and cardamom (*Elettaria cardamomum*) grow from thick, rootlike rhizomes and send their spearlike leaves up to gather the filtered light of the forests followed by tall spikes of light-colored blooms.

Black pepper, the King of Spices, was (and is) a foundational commodity in the international spice trade and a sharp spur for exploration.[2] Its small and wrinkled black fruits traveled the world along both the ancient Silk Road and the oceanic paths of the Spice Routes while inspiring global explorers such as Zheng He, Vasco da Gama, Ferdinand Magellan, and Christopher Columbus. Spice traders transported the climbing plants to Indonesia and Malaysia, where they financed large port cities, and black pepper is currently produced also in China, Indonesia, Vietnam, and Brazil. In the Kampot area of southern Cambodia there grows a specialty type of black pepper that falls under the global category of Geographical Indication, a label that is also applied to other distinct products, including Roquefort cheese from Roquefort-sur-Soulzon, France, Pinggu peaches from the Pinggu district near Beijing, China, Ethiopian coffee, and Darjeeling tea. Although not native to Kampot, pepper plants grow well in the area's quartz-rich soils and produce fruits described as having citrus-fruity notes with jasminelike aspects. Characteristics of place provide terroir to grow recognizable and desirable agricultural products unique to their origin. In India, where pepper originated, vines are planted in gardens at the onset of summer monsoons to be harvested in winter and may be trained up a jackfruit or mango tree.

We call it black pepper, but the plant also gives us the unripe green form and a peeled version called white pepper—products that differ only in the matter of processing, which gives them a slightly different flavor. I hope that you have experienced the amazing burst of sharpness that is freshly ground black pepper. This tang, this bite, this "pepper" comes from the dark coat. Drying in the sun creates a wrinkled black covering from the ripe red fruit while activating pungent volatile oils such as piperine and limonene to provide that wonderful burst of sharpness and citrus you smell as you grind it. If processed to prevent ripening, by heat for example, green berries can be preserved in brine for a more herbal pepper taste. White pepper is somewhat milder and less complex because the dark outer skin has been removed. It may also have a transitory fecal blast of scent—probably from skatole and other volatile chemicals produced by traditional processing in water called retting—that moderates into a uniform sharpness. Indonesia is a large producer of white pepper and considers it to be a value-added form of black pepper due to its versatility and wide range of uses in cooking. What about pink pepper? Sometimes it is the unripe form of black pepper, but there is also pink pepper (*Schinus molle*) that comes from a completely different plant. Native to northern South America, the tree that produces pink peppercorns has fragrant flowers and may be grown as an ornamental. In Florida the related Brazilian pepper, *S. terebinthifolius,* is an invasive tree that also produces dark pink berries for birds to eat and disperse.

Peppers are useful in perfumery: a nicely distilled black pepper essential oil can add a perfect sharpness modified by citrus notes that segues into an elegant wood aspect. Green pepper essential oil, like the spice, has a lovely unique greenness on top of the sharp, and white pepper in tiny doses adds a musky interest. Pink pepper has a dry and woody spiciness that is also somewhat fresh, and it is beautiful blended with lavender for a top note. Piperine from the coat of black pepper and its synthetic relative picaridin are both distasteful to insects, including mosquitoes, and may be used in natural insect repellents.[3]

From earliest days the tiny seeds of *Piper nigrum* likely made their way along travel routes throughout Asia, reaching Egypt, where black pepper was used in embalming, and China, where it was considered medicine. The Assyrians and Babylonians were trading pepper, cardamom, and cinnamon from the Malabar Coast as early as 2000 BC. Pepper vines were transported to other tropical areas such as Sumatra,

Java, and the East Indies as the spice trade progressed. Well known in ancient Rome, pepper was kept in warehouses called *horrea pipera-taria* and was demanded as a ransom by Alaric the Visigoth king when he conquered Rome in the year 410. In England, black pepper was at one time used to pay rent (the so-called peppercorn rent), and one of the oldest guilds in the City of London was the Pepperers' Guild, registered as wholesalers in 1328. Workers in the pepper industry were often required to cut out their pockets and sew the holes shut to prevent theft of the highly valuable peppercorns.

Like many people in the West, I was familiar with the travels of da Gama, Columbus, and Magellan—at least in a superficial way—before beginning the research for this book. The story of Zheng He (c. 1371–1433), however, showed me a new and fascinating side of early exploration. This renowned explorer was born Ma He to Chinese Muslim parents and was captured at age ten by the first Ming emperor of China, Zhu Yuanzhang. Castrated at age thirteen, he grew into a large and fearsome warrior and, taking the name Zheng He, became superintendent of the office of eunuchs. Under the third Ming emperor, Zhu Di, he commanded the Treasure Ships, a fleet of massive vessels built for exploration and transporting treasure finds. Zheng's fleet sailed from southern China past Sumatra and Malacca to Sri Lanka and the Malabar Coast of India in a symbolic journey of diplomacy and military might. Although they traded in black pepper for the Chinese market, the ships of the treasure fleet are known more for transporting such wonders as African giraffes and the first eyeglasses. China withdrew from world trade not long after Zheng's voyages, just sixty-five years before Vasco da Gama first sailed, leaving room in the oceans for European explorers.[4]

Pepper continued to be the foundation for trade in the centuries that followed, more by reason of pure volume than per-pound value, and supported the wealthy port cities of Europe. Once the vines reached Calicut (modern-day Kolkata), Sumatra, and Malacca, they were grown there on forested hillsides. At harvest time in the winter locals would collect the seeds, load up their small boats, and sail downstream to emerge quietly from the forests to meet the traders. As I read stories of millions of pounds of pepper originating in various Asian countries during early trade, and imports to Rome by the ton, I have a vision of a virtual waterfall of the tiny black orbs flowing out of tropical forests and creating dark currents over the oceans.

As with many other precious spices, there were tales of dangerous creatures and conditions associated with the mysterious harvest. Marco Polo, originator of many tall tales related to spices of the East, spoke of forests where black pepper vines grew surrounded by dangerous snakes that had to be burned out by fires that also created the dark wrinkled fruits. Some legends have a bit of truth. Research done in this century is bringing attention to the dangers of snakebites, especially by venomous snakes, worldwide. In Malabar, snakebites occur most frequently among male patients aged twenty-one to forty during August, September, and October. Although the vocation of the patients and circumstances of the bites are not recorded, these patterns describe potential agricultural workers. And they are scary snakes, like cobras and vipers and the banded krait, that lurk within the lush vegetation.[5]

Plants in the *Piper* genus have a spike of tiny flowers that open over a period of several days and may be pollinated by insects as well as by monsoon winds and rains that disperse pollen. As a vine, the pepper plant requires support and thrives in the moist and shady conditions of the forest understory. Picture a tropical or semitropical forest, warm and humid, with layers of vegetation from the tallest emergent trees towering over tall shrubs and intermediate trees in the understory, abundant vines and lianas growing up the trunks, and a covering of herbaceous plants low to the ground. When a pepper vine finds the right conditions for growth, it will climb toward the light while attaching its roots to the bark of a supporting trunk. This is the pepper vine's ancestral home, and with the right moisture at the right time, with perhaps some monsoon winds, visits by small insects, and dewdrops in the morning to assist in pollination, flowers bloom and are pollinated and the pepper fruits develop to produce their store of protective aromatic chemicals. A prolific genus with over one thousand species, *Piper* also contains *P. longum,* or long pepper, *P. cubeba,* or cubeb or Java pepper, and *P. betle,* or betel, which is popular in areas of Asia where they chew the leaves. About seven hundred species grow as herbs, vines, and shrubs in the Neotropics, where they are important understory plants, and fruits are often eaten by bats that also disperse the seeds.[6]

Ginger and cardamom are plants that grow close to the ground under the trees and vines of humid forests. Both are members of the zingiber family, or Zingiberaceae, perennial herbs with rhizomatous roots—basically underground stems that grow both roots and new shoots. Common in wet understory habitats of their native Southeast Asia, zingibers are acclimated to the monsoon climate, and some go dormant during the dry season, losing their leafy aboveground parts. Ginger (*Zingiber officinale*) carries its spice in rhizomes, while cardamom (*Elettaria cardamomum*) sends it to the seeds: both plants have been extensively domesticated and are grown commercially around the world. Members of the ginger family have highly modified flowers, bilaterally symmetrical with a large striped lip in yellow and pink or purple. There are five or six fertile stamens that have diversified into various petal-like organs. Heavily domesticated, ginger no longer produces flowers, unless left unharvested or started with larger rhizome cuttings than usual, but it grows quite nicely from its fleshy and aromatic rhizome.[7]

Both India and China undertook cultivation of ginger early on, and it was carried along on trading ships, perhaps both as food and as antinausea medicine, since the rhizomes were easily grown on board. The story of spices is often a story of medicine first and food later. Ginger has long been known as an antidote to nausea and aid to digestion. As an addition to food, medieval European cooks and diners appreciated spices like ginger to make their bland food and salt-preserved meats palatable. The fresh root is commonly used in a variety of cuisines, as is the powdered form, and in some places, the young stem is also eaten. When fresh, the descriptors zesty, pungent, citrusy, and fresh come to mind, and a nice, fresh ginger essential oil has the same effect. When dried and powdered, ginger loses its freshness and becomes somewhat woody, although still pungent and spicy. The same is true of most ginger essential oils.[8]

Ginger's relative cardamom carries its spice in its seeds. Also native to the Western Ghats, where it grows wild in wet evergreen forests, cardamom has been introduced as a cultivated crop in many countries and grows best in hilly regions between 2,600 and 4,300 feet in elevation. This is the Queen of Spice and considered the third most expensive spice in the world after saffron and vanilla. Cardamom has been used for millennia, according to ancient Sanskrit texts, and is a familiar addition to coffee in some Middle Eastern countries. Like ginger, the plant is a rhizomatous monocot, with leaves between one and three feet long. Flowers grow in long panicles that originate from the base of the plant and may stand upright or lay right down on the soil or forest floor litter, some holding as many as forty-five flowers. Cardamom flowering peaks at the height of the monsoons and generally lasts, in India, from May to October. A large labellum, or lower lip, on the flower provides a space for pollinators to land and pink or violet nectar guides to lead them to nectar, but a pollinator must squeeze between anther and labellum, depositing pollen onto the stigma on entry and picking up pollen as it backs out. Each flower may be visited multiple times and, if fertilized, produces a single capsule with about ten seeds, which are the spice. The subtle and aromatic spice smells first of pine—yes, pine—in both essential oil and freshly ground forms, with a hint of citrus and floral that quickly modifies into a nice spicy, woody finish. Cardamom finds its way into many of the great spice mixtures of the world's cuisine, including Yemeni *zhoug,* Syrian, Turkish, and Iraqi *baharat,* Indian curry powders, chai, and khorma blends, and Malaysian masalas. Cardamom pods were an early trade item for Arab traders as well as a flavoring for coffee used by Bedouins as they gathered around their campfires. The Vikings brought it to Scandinavia, where it features in sweet pastries like those my mother used to make.[9]

Like ginger, cardamom is grown in areas outside its original home in southwest India, and Guatemala is the largest commercial producer today. It is also grown in Sri Lanka, Papua New Guinea, and Tanzania. When cardamom is at home in the Western Ghats, it may be used as an alternative crop in such cultivated forest crops as coffee, but small, isolated pockets of wild cardamom still grow at a distance outside the cultivated areas. Originally, cardamom was regularly harvested from the wild, but a transition toward domestication meant that locals would clear the understory of trees to encourage cardamom growth. As is often the case around the world, demand for more cardamom and easier access led to the loss of multilayered forests and a more simplified structure in the fields, with artificial structures or tree plantations providing the shade.

Most crops need pollinators, and they are often bees of some sort that busily fly about their business of gathering honey and pollen while incidentally providing a grower with seeds for next year's crop. But many spice plants are now grown in areas away from, and with different ecology than, their habitats of origin. Cultivated cardamom flowers have developed longer blooming time and more nectar, which attracts social bees like honey bees and some stingless bees, replacing the native solitary bees that were the original pollinators. Wild carda-mom plants are still pollinated by solitary bees like leafcutters but have a lower density and, not being managed, may be more subject to damage from elephant herds, overgrowth by other vegetation, fallen trees, and forest fires. There are two species of honey bee suitable for pollination on Indian cardamom plantations, *Apis dorsata* and *A. cerana*. *Apis cerana,* an Asian honey bee, can be kept in managed hives, but they may be expensive to rent, and many landowners lack the knowledge to maintain them. *Apis dorsata* bees do not settle in one place but move according to resources and have become rare as traditional honey collectors have devastated their populations. The sting-less bee, *Tetragonula iridipennis* (formerly *Trigona iridipennis*), though smaller and inconspicuous, also pollinates cardamom flowers. Many of the same bees will also pollinate coffee, which is often cultivated with cardamom, leading researchers to recognize the importance of supporting a healthy population of bees. A regular source of food in the form of flowering plants may help keep pollinators on the farm, and floral calendars provide information on bloom sequence to farm-ers that helps maintain a food source year-round for wild pollinators. Adding flowering trees in and around the fields as recommended by a floral calendar may provide shade for both coffee and cardamom, natural mulch, nesting sites for bees, and perhaps a site to grow a few black pepper vines. Forests of the Western Ghats are rich in vegetation and complex in structure, with a climate capable of supporting flow-ering plants and their pollinators year-round.[10]

Spice of the phoenix, without the bite of black pepper but with a homey and slightly sweet bite, cinnamon finds its way into many recipes. This spice comes from the bark of a tall tropical tree whose wood may be used in carving and can have a rosy grain that is slightly lustrous. There are two common species in use, *Cinnamomum verum* (also unofficially called *C. zeylanticum*) and *C. cassia*. Both types show up in Chinese five-spice powder, hot chocolate from Mexico and Central America, Lebanese lamb dishes, and gooey, scrumptious cinnamon rolls in shopping malls around the world. *Cinnamomum verum* grows in the forests of the Western Ghats and Sri Lanka. The best is thought to come from Sri Lanka, where it develops a classic and subtle flavor that is sweet and spicy without being bitter. *Cinnamomum cassia* grows wild in southeastern China, Assam, Myanmar, and Vietnam. It is a bit simpler in flavor, yet intense with a touch of bitter. *Cinnamomum verum* has a single fleshy fruit that is attractive to native fruit-eating birds that disperse it throughout the forest. Various species of cinnamon trees may form an important part of the tree canopy in their tropical habitat, and in some places, such as the Seychelles Islands, they can become dominant and preclude other native trees from growing. As with most spices, the composition of volatile aromatics in cinnamon is complex, consisting of more than seventy different compounds that provide the characteristic taste or smell. In cinnamon it is cinnamaldehyde that provides a sweet and floral cinnamon fragrance, but the spice also contains clovelike and spicy eugenol. Flowers of the cinnamon tree have a cinnamon-floral fragrance and taste of allspice and pepper. Essential oils may be obtained from many parts of the tree and are mostly used in flavoring.[11]

As with other spices originating in Asia, there are legends describing the origin and harvest of cinnamon. Both eagle and phoenix were associated with cinnamon, the phoenix with the heat and dryness of the spice that evoked hot suns of Arabia and the eagle that used it as nesting material. Never mind that cinnamon did not grow in Arabian deserts; myths, and the tendency to classify plants as hot and dry or cold and wet, made phoenix and cinnamon a natural match. The phoenix's nest was built of spices such as cinnamon, myrrh, and frankincense that also created a perfumed pyre at the end of the bird's life before its rebirth. For harvest, according to Herodotus, cinnamon sticks had to be retrieved from the nest material of giant eagles, a task that was accomplished by feeding the eagles large chunks of meat. When the eagles carried the meat to the nests, the weight would cause them to collapse, allowing the cinnamon to be gathered. Or, also according to Herodotus, some trees (possibly cassia-type cinnamon) grew in shallow lakes protected by loud and pesky bats so that harvesters had to don protective leather garments to gather it.

However it is harvested, cinnamon comes from the inner bark of stems and trunks that is peeled off and dried in curled pieces about three inches long, making the distinctive quill. Poorer-quality bark, smaller pieces, and leavings, or quillings, may be ground up for sale or distilled to yield an essential oil. Cinnamon was commonly used after death to preserve or commemorate ancient Romans and medieval Europeans of importance and wealth, including the dictator Sulla, who had an effigy created from cinnamon. Other emperors simply ensured that cinnamon was added, along with other aromatic spices, to their funeral pyres—perhaps to reenact the rebirth of the phoenix and the triumph of life over death. A Franciscan monk of the fourteenth century, Juan Gil de Zamora, provided recipes to cure raptors in his scientific encyclopedia. To cure a goshawk of a headache, he recommended a mix of cloves, cinnamon, ginger, pepper, cumin, salt, and aloes. If you still have your fingers after that treatment, you may want to give your rheumy falcon a blend of ground amber, ginger, and pepper. For humans, cinnamon is one of the better-known medicinal spices and was considered a heating spice and, according to Galen's book *Concerning Antidotes,* was therefore an antidote to the dangerously cooling properties of hemlock.[12]

Nutmeg and mace (Myristica fragrans) *with clove*
(Syzygium aromaticum)

4

The Spice Islands

Once upon a time, sailors to the Spice Islands of the Moluccas (Maluku) knew how to get some of the precious cloves that grew there. According to legend, upon arriving at a local beach they would leave trade goods in a heap and then return to the ship. The next morning, they would return to find a pile of cloves at that spot in an amount the locals deemed appropriate. Whether true or not, this simple story of peaceful trade, respect for native people, and value for value did not last. These small volcanic islands in the Indonesian archipelago and their iconic spices became central players in a contest for wealth and power that played out across the globe. Scattered in the ocean between Australia to the south and Asia to the north, and bordered by Papua New Guinea and Malaysia on the east and the west, the mountainous islands of Maluku and Banda vary from tiny to, well, not so tiny and make up the Spice Islands.[1] They occupy a sort of sweet spot between Australia and Asia that has led to the evolution of a diverse and unique flora and fauna, including nutmeg trees (*Myristica fragrans*), which are tall, evergreen, and dioecious (meaning there are both male and female trees). A nutmeg tree takes up to seven years to flower before growers can distinguish between male and female and cull out male trees, leaving just a few to fertilize the seed-producing female flowers. Nutmeg is the large seed of the trees. Outside its native habitat in the Banda Islands, nutmeg is grown in Grenada, India, other parts of Indonesia, Mauritius, Singapore, South Africa, Sri

Lanka, and the United States. Nutmeg is all about the terpenes that help to protect against insect damage—sabinene, which is woody, warm, citrusy, and alpha and beta pinene, which give a bright and friendly flavor. Myristicin is the characteristic aromatic constituent that gives a warm and slightly sweet woody note to the fragrance. A second spice, mace, is derived from the lacy, threadlike covering of the nutmeg seed and may be used in some of the same dishes, but it is also somewhat more complex in its fragrance. Many have noticed that nutmeg seems to influence mood, being generally elevating in nature but thought by some to be hallucinogenic in high doses.

Canoes were likely the first oceanic vessels to approach the isolated islands where nutmeg grows. They would have passed by abundant sea life, including turtles, whales, and vast populations of fish on coral reefs, and felt the oceanic winds blow over while warm rains fell on the eleven small islands of the Banda archipelago of Indonesia. Perhaps the winds carried the fragrance of the tall evergreen nutmeg trees to natives who fished in clear blue waters. Arab traders came next, and then the Portuguese, British, and Dutch. The story of trade in the Banda Islands exemplifies the lengths to which companies would go to monopolize trade in highly desired items. Ships in Magellan's Portuguese fleet visited in 1521, followed by Dutch traders who established the Dutch East India Company, which brutally controlled most of the islands that grew the nutmeg trees. Through slaughter, expulsion, and slavery they came to dominate the archipelago and hence trade in nutmeg, mace, and cloves. By the seventeenth century the Dutch lacked just one island for total control, Run (Rhun), held by the British. However, the Dutch owned the island of New Amsterdam in the American colonies. Swapping New Amsterdam, which became known as Manhattan, for absolute control of nutmeg along with a sugar interest in South America proved mutually agreeable, and the Treaty of Breda was signed in 1667. The Dutch built forts and trading centers on the islands and controlled trade for more than three hundred years, until Indonesia fought for and won independence following World War II.

The Dutch East India Company was interested in the botany of places that gave rise to their spices, such as medicinal plants that could treat unusual tropical diseases not amenable to the traditional European simples. One of their employees, the botanist Georg Everard Rumphius, lived on the island of Ambon from 1653 until his death in 1702 where he made detailed descriptions of native plants in the form of notes and drawings. The nutmeg tree, for example, he described as "handsomely shaped and glossy-leaved," with an oval fruit reminiscent of a peach. He referred to its use by the natives to assist men in pleasing women and prostitutes to make merry but also described it as a mood enhancer. Rumphius's copious writings have been compiled into a beautifully illustrated six-volume set of books, *The Ambonese Herbal,* which provide a comprehensive look into the botany, ecology, and anthropology of the Spice Islands.[2]

Yielding both the nut and the fragrant covering that gives us mace, nutmeg flowers are pollinated by thrips and small beetles, with beetles seeming to be the most effective. The term *cantharophily* means "beetle love," and it specifically refers to a process of pollination by beetles in certain plants. For some, the showiest of which are magnolias and lotuses, beetles have evolved as the primary pollinators. In cantharophily, scent may be the first attractant and has been characterized variously as strongly fruity, fermented, and slightly spicy, with tones of lily of the valley, but some flowers will smell of decay or sweat. Once they approach the flower, beetles respond to the signals of color and shape, landing within the large, curved petals and seeking food in the form of pollen. Some cantharophilous plants are even able to warm the flowers at night, attracting beetles to shelter within the petals and enhancing the release of scent. Because many beetles are poor fliers and are happy to sit where they land, they carry out other life functions such as eating, defecating, and mating while ensconced within the flowers. This has led to the term *mess-and-soil pollinators* to describe their actions. These often hairy beetles will chew flower parts, rampage around while mating, sometimes leave an incubating brood behind, and generally create havoc and damage within the flowers.

Occasionally a beetle will be an effective and faithful pollinator for certain flowers, but often it is hit-and-miss as a beetle flies about among various flowers. Some cantharophilous flowers have taken on the role of good host, not only providing warmth and shelter but also exhibiting specific food bodies, specialized tissues for the beetles to munch on to distract them from eating petals. Some members of the nutmeg genus, *Myristica insipida* specifically but probably also *M. fragrans* (which has a nearly identical flower), have a slightly different relationship with their tiny beetle pollinators. The behavior is called microcantharophily by some scientists, and these beetles seem to be somewhat better behaved. Nutmeg flowers are small, fragrant, and massed in bunches rather than being large and showy like typical cantharophilous flowers. They also bloom at night when beetles are not necessarily flying about. Around dawn, however, beetles will find fragrant male flowers, feed on pollen, and gather a few grains on their bodies. Their visits are short and limited to eating pollen—no messing and soiling. Female flowers provide no pollen reward but attract pollen-laden beetles by mimicking male flowers in scent and appearance.[3]

Cloves (*Syzygium aromaticum*) are the hardened and unopened flower buds of a lovely evergreen tree that may live up to 150 years, bearing a bumper crop every four years or so. The aroma of cloves can be overpowering. It is strongly aromatic and spicy, with cinnamon overtones and a dry woody aspect. Eugenol adds a medicinal and warming aspect to cloves and is produced as the bud matures. Clove leaves are also aromatic and may be distilled for their essential oils but are a secondary product and are not usually picked since that may reduce flowering. If allowed to bloom, the flowers are crimson in color with an abundance of stamens. Clove trees can be sensitive to broken branches and damage, so buds must be carefully plucked by agile fingers and dried in the sun to bring out their aromatic constituents. As highly useful spices, strongly scented, and small, cloves likely traveled with many a trader and perhaps changed hands and pockets numerous times as they traveled by land and sea with Arab traders who kept their sources a closely guarded secret. Ancient Egyptians used the spice, Chinese beginning as early as 200 BC were known to use it, and Crusaders and Romans alike were big fans. In Indonesia, kretek cigarettes made with cloves with their fragrant, crackling smoke are popular. Cloves have been used to sweeten the breath and create both perfume and incense.[4]

Magellan almost made it to the clove island of Tidore but was killed in the Philippines on the way. Two of the five original vessels in his fleet made it to Tidore and filled their hulls with a dangerously huge load of cloves. The only vessel to return to Spain was the *Victoria,* under the command of Juan Sebastián Elcano, who paid for the expedition with the profit from cloves in his hold and in return received a lifelong pension as well as a spicy coat of arms with two cinnamon sticks, three nutmegs, and a dozen cloves. The Portuguese were unpopular conquerors, but the Dutch who followed were merciless. They cleared the islands of Tidore and Ternate to plant clove trees in Ambon, where they could control both prices and traders. Now cloves are grown in other parts of Indonesia, as well as in Madagascar, India, Tanzania, and Brazil.

When I peruse literature and books with lists of aromatics in spices, the primary descriptor I find is "spicy." If I search on the Internet for definitions of *spicy,* I turn up synonyms like piquant, fragrant, tangy, peppery, hot, seasoned, pungent, sharp, and flavorful. Merriam-Webster's online dictionary tells me that the first use of the word *spicy* was in 1562 and reminds me that alternate meanings include lively, salacious, scandalous, and racy. The word *spice* comes from the Latin *species,* which means "sort or kind," which the French shortened to *espice,* and which then became *spice* in Old English. Referring to a fragrance (or taste) simply as spicy does not do much to distinguish among the spices in this section, and other descriptors can help to group them. Remember that spices, and most aromatic plants, produce and contain volatile chemicals by the hundreds, and a few of these may be major contributors to flavor or fragrance, many others act as subtle modifiers. Cooks can generally make these distinctions, as can perfumers, so I put on my perfumer hat and gave it a go in the interest of finding a way to describe the scent of different spices.[5] There is no right answer when sniffing and describing, so spend a bit of time with your favorite spices and see what words come to mind.

In addition to being spicy, cinnamon and cloves are both warming and woody, and they contain β-caryophyllene—a constituent that is described as spicy, woody, and dry. When I grind cinnamon and cloves, I find dryness in the cinnamon along with a sweet and uncomplicated fragrance of fresh woodiness. Cloves are also dry in fragrance but have a bit of green as well, which gives them a hint of stemmy carnations-in-the dirt-fragrance. Beta-caryophyllene is useful in plants since it attracts ladybird beetles to prey on the aphids that may eat tasty leaves. It is found in a variety of herbs and spices and is also one of the terpenes in cannabis.

Nutmeg and mace have a medicinal hint touched with sweet and floral. In addition to the cinnamon note of eugenol, they have geraniol, which adds a floral freshness, and cineole, which gives that eucalyptus or medicinal aspect. When I grate a large nutmeg seed, I find that the sharpness dominates at first, but if I give my nose a moment, the sharpness is followed by slightly sweet freshness. Eugenol is found in a variety of plants and is both attractive and repellent to insects, depending on where it occurs. Fruit flies especially love the molecule: males will ingest fragrances containing eugenol to modify it for use as a pheromone, and so some flowers produce eugenol to attract fruit flies as pollinators. Stored in the rectal gland of the tiny flies, eugenol will be modified internally and released later to attract females.

Rather than spicy, black pepper is better described as hot or pungent—a straightforward effect due to piperine—but it also has woody, lemony, and piney notes from pinene and limonene. When I find a good essential oil of black pepper it has a very transitory top note consisting of piney combined with sharp piperine, just like freshly ground peppercorns, but there is also an elegant woody drydown.

Cardamom has a complex scent with a bit of linalool that adds floral to the medicinal. Grinding it fresh is a revelation to me of how sweetness and turpentine, lemon and floral, can produce almost a perfume in itself. When I write my evaluation the words that come to mind are complex and pointy, ethereal with something like precious leather, and all aspects perfectly balanced. Linalool has a floral fragrance with woody undertones and is widely present in plants with varied effects—it may repel herbivores or attract their predators, but it may also contribute to the perfume that gathers moths to a flower.

Fresh ginger is quite different from the dried and powdered we might find on our spice shelf. The defining aroma of fresh ginger is provided by gingerol, which delivers the characteristic bite. When dried or heated gingerol may be converted into zingerone, which has a sweet and spicy vanilla fragrance along with the bite. The fragrance of these molecules is modified with floral from geraniol and linalool as well as a touch of medicinal cineole. My powder, although a bit old, has the sweetness and woodiness I have come to expect.

Saffron (Crocus sativus) *flowers*

5

Saffron, Vanilla, and Chocolate

The precious fine, dark orange-red strands of saffron originate not from tropical forests but from rocky soils in temperate areas like the Kashmir Valley in the Jammu and Kashmir state of India and in the Mediterranean basin, where the plant has been grown for more than a thousand years. In fall, small, deep lavender flowers grow from bulblike corms buried in the soil, sending up a few spear-shaped leaves and a flower that arises directly from the base. This tender cup-shaped flower bears in the center three deep red stigmas that will become the most valuable spice in the world. The flower is *Crocus sativus,* and at one time it covered the valleys of Kashmir—at least according to legend. One autumn when Alexander the Great reached Kashmir he had his army camp in the valley for the night: when they awoke in the morning they were surrounded by a sea of violet flowers. These same flowers held a deep fascination for Greeks during the Bronze Age and are depicted in statues and drawings, whether being gathered by skirted women or as a simple floral decoration. Frescos on the walls of the Minos Palace in Crete depict saffron gatherers from as early as 1700–1600 BC.

Saffron is a triploid plant, which means it is sterile and needs to reproduce vegetatively (without exchange of gametes) from divisions of the expanded underground stem, called a corm. Saffron likely arose from a closely related plant, *Crocus cartwrightianus,* but at some point, the mutated form we call *C. sativus* appeared with its longer and more obvious stigmas.[1] Unique for both color and its deep, earthy, and musky flavor, saffron spice is treasured around the world as an additive to foods but has also been used in medicine as well as in paints and dyes. With hints of grass, earth, honey, sweet floral, and bitter, tiny bits may transform a meal.

Saffron crocuses bloom in autumn, unlike the well-known garden crocuses familiar to many that are harbingers of spring. Flowers are gathered by hand, stigmas are separated from the flowers, also by hand, soon after picking and are dried in the shade for flavor to develop. Saffron is grown in Iran, Spain, India, Greece, Argentina, and the United States, where over seventy thousand individual flowers produce one pound of the dried spice. This translates to potentially millions of the petite plants growing throughout saffron-producing countries. One of the oldest saffron growing areas is in old lakebeds of Kashmir, where climate change is affecting patterns of rainfall and daily temperatures to the detriment of production and quality of the valuable spice.

Although *Crocus sativus* does not reproduce sexually, saffron crocuses owe the appearance of their stigmas to pollinated ancestors. Crocus flowers of a variety of species open in early spring or in fall when lower temperatures and rain pose challenges for both flower and pollinator, but these flowers have an answer for that. Their little floral bowls on delicate stems can track sunlight, reflect it inward, and raise the temperature inside the flower by a few degrees; some species close their petals to retain moisture and protect pollen. This strategy is present in some other geophytes—plants with underground structures like bulbs or corms—that have bowl-shaped flowers in the red, pink, white, and purple range. They often originate in climates with cold or cool spring and fall weather, such as Mediterranean, montane, and subalpine habitats, and include anemones, buttercups, and primroses, as well as crocuses. Flowers like this are called microgreenhouse flowers: they have reflective inner tepal surfaces and the ability to store heat in large, darker-colored sex organs such as the stigma and style. The flower's shape and color allow it to act as a parabolic mirror and reflect heat energy. In early spring and fall, potential pollinators, a few species of solitary bees and syrphid flies, are small in number compared to those available during the peak of spring flowering. It seems to me that the blankets of deep purple flowers with deep red stigmas on the stark autumn landscape would provide an attractive signal to roaming bees seeking a warm welcome. Warming of the flowers may also benefit pollen development and pollen tube growth.[2]

To attain the deep reddish-orange color in the stigma, crocuses produce pigments called carotenoids that may be modified into fragrant compounds including ionones that have a woody, violet, floral scent and are an aroma constituent in tea, in quite a few fruits, including grapes and berries, and in roses, tobacco, and wine. Carotenoids give plants such as tomatoes their orange and red colors and may be modified to create vitamins, including vitamin A. In *Crocus sativus,* other secondary compounds include crocin and crocetin for color, picrocrocin for taste, and safranal for fragrance that develop as a result of drying and enzymatic action on the carotenoids.[3] Once dried, the tiny stigmas are carefully sorted, and intact ones are packaged for sale at high-end grocery stores. A small part of the saffron harvest goes to produce a solvent-extracted absolute, a perfectly gorgeous and unique perfume ingredient in which the haylike earthiness of the spice becomes deep and leathery while retaining the characteristic bitter herbal nuances. The absolute is used in perfumery for a leathery note and in many Indian attar-based fragrances.

Since the rich orange-red of the stigmas turns deep yellow in food, the word *saffron* has come to describe a color of the same hue and the robes of Buddhist monks. Dye used for robes was in fact more likely turmeric, given the remarkably high cost of saffron and the humble nature of the monks. Saffron was used as a paint to achieve a less expensive gold leaf effect in some medieval European manuscripts, and it did indeed produce a glowing and rich yellow color. In her beautiful book on color, Victoria Finlay describes the color of a few strands of saffron infused overnight into egg whites as luminous, as if the whites had reclaimed their yolks. Wealthy patrons of medieval monasteries would often donate items to memorialize themselves in both life and death. Smelling, seeing, touching, and even eating donated items would remind monastics to pray for both living and dead donors. Saffron, with its bold color, aroma of luxury, and unique taste was a multisensory reminder of the giver.[4]

Chocolate and vanilla: most of us cannot think of one without the other. Fortunately for early inhabitants of Central America and Mexico, vanilla and cacao beans could be found growing there, although not exactly next to each other. Take some cocoa, a touch of vanilla, add in some chili for zip, and a few sprinkles of cinnamon, and you have my very favorite chocolate combination. Mayas and Aztecs perfected the art of blending chocolate and vanilla with herbs and spices to create a somewhat bitter concoction; sugar was to come later, thanks mainly to Europeans. Mayas, as early as fifteen hundred years ago, may have pounded and cured cacao beans, grinding them into a paste that would harden into blocks. Bits would be broken off and added to hot water as desired. Or ground corn, chili peppers, and vanilla would be served to Aztec gods in the drink called *xocoatl* that was also used to combat fatigue. Neither chocolate nor vanilla grew in the higher and drier habitats where the royal Aztec courts were established, but chocolate, at least, was an item of trade with lowland villages of southeastern Mexico.[5]

The first European to taste chocolate was likely the Spanish conquistador Hernán Cortés when he encountered Montezuma II, Aztec ruler, in November 1519. Drawings done by local Aztec botanists of the *tlilxochitl,* or black flower, as vanilla was called, appear not long after in the Badianus Codex, a product of the College of Santa Cruz in Mexico City.[6] We do not know who was the first to follow their nose, brushing away foraging ants to find sweetly scented vanilla pods drying on the forest floor in Mesoamerica, but we do know that vanilla was recognized early on as the perfect partner for cocoa and that beans of the vanilla orchid were transported to Spain somewhere in the mid-1500s. Vanilla made it to the rest of Europe and then to the newly created United States in a convoluted return journey overseas, not from its neighbor to the south. The beans quickly became popular and were featured in a recipe for vanilla ice cream by Thomas Jefferson, who learned about it in France.

Vanilla (*Vanilla* spp.) is the fruit of orchids. Gangly and wrinkled, the dark brown beans are produced from graceful orchid flowers on vines suspended above a tropical forest floor. For many, the term *vanilla* means bland and plain, not particularly interesting. Which is really and truly not the case. If you have sniffed a really good bean, nicely cured and still a bit bendy, you know that it has aspects of earth along with sweetness, darkness to go with floral, and more complexity than those little dots in your vanilla ice cream seem to indicate. If you have been lucky enough to find beans from around the world you can sample the earthy tones of Ugandan beans or the floral and complex heliotrope notes of a beautiful *Vanilla* × *tahitensis* hybrid from Tahiti. Mexican beans, *V. pompona,* are dark, winelike, and strong, so they work well with chocolate, cinnamon, and coffee-flavored liqueurs. Most of us are more familiar with the beans produced in Madagascar known as Bourbon vanilla, *V. planifolia,* that have creamy notes of hay and brown sugar.

Early vanilla harvests in Mexico and Central America were likely more a matter of gathering ripe vanilla pods from the woods and possibly transporting a few vines to grow around the villages. Because vanilla vines are so easy to propagate vegetatively—basically by cutting bits off the vine and putting them next to a support tree, adding a bit of fertilizer, and providing some shade—there seems to have been little done in the way of bringing in new genetic material for commercial crops. Selective breeding with different types would perhaps help improve disease resistance, increase yield, or produce new varieties. In a family with more than a hundred species it seems likely that there are more kinds of vanilla out there to be found. Even within the commercial species, *Vanilla planifolia,* there are types with variegated leaves, as well as one with drooping leaves like the ears of a donkey called *oreja de burro.* Beans from Réunion Island produce what we call Bourbon vanilla. *Vanilla odorata* is grown in Latin America and has nicely fragranced pods that may be used to flavor rum. It may also be one of the parents of the Tahiti vanilla. A proposed new species from Costa Rica, named *V. sotoarenasii,* has smaller, distinctively shaped leaves and beans with fruity-almond and licorice notes.

The vanilla group of orchids is an ancient lineage, and species can be found in tropical forests on five continents.[7] They are nearly all vines, and some can attain a huge size as they branch and spread throughout a forest, clambering up trees and showing off large springtime flowers in shades of green, white, yellow, or purple in various combinations. Petals form a tube around the sex organs: both female stigma and male pollen-producing anther occur on the same flower but are separated by a membrane called a rostellum that prevents self-fertilization. Ancestral vanilla likely comes from semi-evergreen forests of less than three thousand feet above sea level in the area around Oaxaca, Mexico. Commercial vanilla, most of which is cultivated outside its native habitat, does not have local pollinators, requiring humans to perform the task, which must be completed in the hours just before and after dawn—preferably humans with small and nimble fingers, like the young enslaved worker Edmond Albius, who in 1841 in Réunion discovered a quick(er) way to hand-pollinate the flowers. Where vanilla is native, local insect pollinators can enter the floral tube, where they become trapped by stiff bristles lining the tube that make entering easier than leaving. In their struggles to leave, the bees pick up a little packet of pollen from the anther and fly off with it to the next flower. Here they deposit their load of pollen on the female stigma during their struggles, completing cross-fertilization. After being pollinated, fruits take about nine months to mature and must be guarded as the beans are cured to produce the characteristic fragrance and flavor material called vanillin. Vanilla pods start out green but split and turn brown as they ripen, becoming rich in oils mixed with a gelatinous substance. Beans that fall to the ground cure naturally and likely attract a variety of insects responding to the scent of vanillin and the tiny oil-covered seeds within.

Vanillin, a product of enzyme action in the ripening bean, provides the characteristic flavor and scent of vanilla. It is produced within in the beans as an unscented compound called glucovanillin and, stored separately, is the enzyme responsible for turning glucovanillin into fragrant vanillin. Vanillin in high concentrations is toxic to living cells and so cannot be allowed to accumulate, thus is stored in the more benign form and created enzymatically after the bean is ripe. As seedpods die, either naturally or through heat and chemical changes, dividing walls break down allowing enzyme and substrate (glucovanillin) to come into contact. Time and heat do the rest. In the commercial process beans must be killed, perhaps in the traditional manner by laying them out in the sun or dipping them in hot water, but they may also be frozen or killed chemically. Sweating follows killing and allows the fragrance to develop. In some traditional methods killed beans may be laid out in the sun on dark blankets that are wrapped around them at night to absorb excess water. In addition to vanillin, there may be as many as 250 aromatic constituents in a ripe bean. Natural vanilla products come from the intensive efforts of tens of thousands of small farmers worldwide who laboriously fertilize, harvest, age, and pack up the beans to sell to a broker who passes the beans to exporters followed by wholesalers and on to manufacturers of extract or powder and finally to the shelves of the supermarket. Those small growers need to pollinate each fruit by hand (up to three thousand a day), wait nine months for the fruits to mature while guarding them against thievery, and then carefully and patiently assist the beans in ripening to achieve their taste and fragrance. These gloriously scented beans account for a small percentage, less than 5 percent, of the vanilla flavoring in our foods and drinks. Most comes from synthesized vanillin.[8]

Vanillin may be commercially produced in a lab or a vat via fermentation by yeast or by chemical reaction. Vanillin was one of the first scented molecules to be synthesized and was created in the lab from conifers. It is still made from leftovers of the wood pulp industry or from rice, some conifers, and cloves—specifically the eugenol in cloves. Since vanillin is a single molecule, it has less complexity and depth than vanilla derived from the plant with its many other aromatic constituents, but that may be desirable in the mass production of both food and scent. Vanilla from the bean is hugely expensive and, like any good agricultural product, varies between years and with terroir. A manufacturer or consumer may not want complexity but instead want their vanilla cookie or gourmand perfume to taste or smell exactly the same each and every time they buy it. Cost and replicability are key to a successful product, and the mass-produced vanillin molecule is the affordable workhorse of vanilla fragrances and flavors. Also, there simply are not enough vanilla beans being grown to meet the demand for this highly popular ingredient. Vanilla beans, despite the facile dismissal of "vanilla" as a descriptor, produce deepness and richness and complexity. Their reflection of terroir and the complex aromas that accumulate as they age at the hands of farmers makes them worth every dollar to many a cook and perfumer.[9]

Cacao (*Theobroma cacao*) is the food of the gods and comes from tropical forests of Mexico and Central America. While Mesoamericans liked their cacao spicy, Europeans found a sweet match made in heaven with vanilla and cocoa. Although both plants grow in humid, tropical areas, cacao probably originated somewhere in the Amazon basin and made its way north to Mexico and Central America, probably with human help, where it has been cultivated for more than fifteen hundred years and possibly up to four thousand years. As with other spices, the seeds are highly transportable, so perhaps cacao made its way along with ancient Americans moving between the two areas before there was any formalized trade between them. The flavorful and aromatic pulp of the fruit was also favored and used to make fermented beverages. The short tree, growing up to about twenty-six feet tall, is an understory component of low-elevation tropical rain forest and is at home growing among taller tropical crop species. White and pink flowers grow right out of the trunk year-round, a trait called cauliflory. Small and lovely, the flowers are grouped in clusters and are visited by a variety of insects but depend on tiny midges for pollination. Not all flowers are pollinated successfully, about 5 percent in the wild, but perhaps just a few are enough since the process from flower to huge fruit is dramatic, requiring a large commitment of resources on the part of the tree. The cacao pod is an exceptionally large fruit, up to ten inches long, containing thirty to forty seeds surrounded by a fleshy white pulp. The pulp is tasty and attractive to such primates as chimpanzees (*Pan troglodytes verus*) in West Africa, Mexican black howler monkeys (*Alouatta pigra*) in Chiapas, Mexico, and yellow-breasted capuchins (*Sapajus xanthosternos*) in Brazil, as well as humans.

Such large fruits would have been a perfect size for extinct megafauna that once roamed the tropical forests and likely played an important role in the original distribution of cacao and other large-fruited plants, along with smaller humans and other primates.[10]

Theobromine and caffeine are found in cacao and are both alkaloids, a family of compounds that add bitterness to a fruit, seed, leaf, or flower and are likely defensive in nature. In both coffee and chocolate, bitter seeds are surrounded by a sweet fruit, luring a seedeater to enjoy the fruit but reject the seeds. Primates that eat cacao fruits will either spit out the seeds or swallow them whole, either way giving the seed a chance to germinate and produce a new cacao tree. If the seed is swallowed it gets a free ride and some nice fertilizer at the end of the trip—preferably in a place conducive to growing up to be a brand-new cacao tree. Maybe under a nice tree canopy in old forest or open secondary forest and perhaps even along a small river course. In a study in West Africa, one of the small plantation owners recognized that local chimpanzees were responsible for "planting" some of his cacao trees. However, he claimed ownership of the trees since he was responsible for their care as he maintained canopy cover and cleared the understory.

But back to the tiny midge. Yes, those irritating, biting, buzzing insects are the most successful pollinators of cacao trees. In addition to requiring a blood meal, female midges visit the flowers to harvest pollen as a protein-rich food to supplement sugary nectar. Other small flies also visit the flowers, but only midges fit the exact configuration of the flower and can work their way through the complicated center to reach the nectar and incidentally pick up a little packet of pollen. When the cacao tree blooms, a midge may follow a light fragrance to the flower. Once there, it lands on the infertile staminodes, several slim, vertical, brightly colored structures that converge like the supports of a tent to form a narrow entrance just the right size for the midge to squeeze through. As the midge moves toward the center of the flower guided by colorful lines and hooded petals, pollen grains from the fertile stamen attach to its body, remaining there as it flies off to make its pollen deposit at the next cacao flower. These tiny midges, less than twelve-hundredths of an inch in length, must carry thirty-five pollen grains (yes, someone counted) to successfully fertilize a flower and cause the large cacao pod to develop.[11]

Since cacao flowers last just a day or two, a local supply of pollinators is crucial, and it has been found that midges, specifically biting midges, need to be in place to carry out this important job. Instead of a sterile, full-sun monoculture with a clean understory, midges prefer a slightly shadier and messier plantation, and they do not fly very far. Water-holding bromeliads and stems of banana plants provide breeding habitat that is cool and shady with rotting leaf litter and pockets of moisture. In Ghana, scientists found that the presence of banana and plantain plants near or within small cacao farms provided habitat for midges and increased fruit set. In farms and plantations where midge habitat was part of the landscape, midges increased dramatically, as did fruit set—although researchers did not mention the nuisance factor of these little biters.[12] Knowing the pollinator helps growers preserve or create habitat that will support it, midges are a small price to pay for the chance for a larger crop. As one of the thirteen most important commercial crops in the world, cacao, like vanilla, depends on small farmers. More than 90 percent of production comes from small-scale farms of 7.5 acres or less that are worked by farming families and yet constitute almost 25 million acres worldwide. This means that cacao provides some form of income to about fourteen million people around the world.

Recent research has found hope for some species of birds that inhabit the fields and wooded areas nearby. A study of overwintering birds found tiny migratory wood warblers (*Phylloscopus sibilatrix*) in various stages of successfully molting their feathers. That the birds were molting meant that the matrix of cacao field and nearby blocks of forest provided habitat with enough food for them to undergo the energy-intensive process of growing new feathers.[13] This bit of good news is, however, in the context of large loss of native rain forest cover to cacao production and the loss of some important bird community members like insectivores, ant followers, and forest specialists.

Fruit of the cacao tree is large, wrinkled, and oval in shape and includes beans, pulp, and the shell. Seeds consist mainly of fat—the useful and beloved cocoa butter—and contain alkaloids such as theobromine. Related species include *Theobroma bicolor,* with the highest theobromine content in the hull of the fruit, followed by flowers and leaves, while caffeine shows up in both flowers and seeds, and another species, *T. angustifolium,* which produces the highest concentration of theobromine in the flowers. More important than the caffeine, at least for some of us, is the flavor that develops during fermentation, drying, and roasting. Genotype, terroir, and fermentation produce different flavors but there are two general classifications of cocoa— bulk (mostly produced in West Africa) and fine or flavor types (from Latin America), which develop fruity or floral notes. Precursors of fine flavor notes likely develop in the pulp of the fruit and permeate into seeds during fermentation, adding constituents such as linalool, myrcene, and ocimene that overlay floral and spicy aspects on top of the characteristic chocolate flavor of the bean. As with vanilla, the bean must die to develop its flavor. Fermentation occurs as a result of a succession of different yeast and bacteria species: as tiny yeast cells digest and ferment the fruity pulp, the bean develops precursor flavor molecules that will be changed on roasting to provide the rich and unique flavor of chocolate.[14]

In addition to processing practices, cultivar and environmental conditions contribute to flavor and quality. Wherever cacao came from, three dominant groups are recognized: Trinitario, Forastero, and Criollo. Criollo is a type that the Mayas cultivated in pre-Columbian times and contains fruity and floral notes. The Forastero type comes from farther south in areas around the Amazon and is generally referred to as bulk chocolate; it is currently grown in Africa, Central America, and Southeast Asia. Although Criollo chocolate is considered to be of higher quality than Forastero, it is also less vigorous and disease resistant. The third type, Trinitario, is a hybrid of Forastero and Criollo types, has a fine aroma, and is more vigorous. Taste tests after curing described Criollo as having complex floral, fruity, and woody aroma notes, while Trinitario cacao was fruitier and greener with woody notes and Forastero was floral and sweet.[15]

Imagine a world without spice, without the complex flavors of an Indian curry, a spicy Middle Eastern kebab, Chinese five-spice powder, or a homemade apple pie with nutmeg and cinnamon. Then imagine a world without chocolate, which is perfect for dessert but goes to another level when used in Mexican moles, and vanilla certainly adds to the richness of the accompanying dessert flan. These spices connect us just as traders by land and sea stitched together the oceans and overland trade routes. Whether it is rediscovering the cooking of our ancestors, sharing recipes with friends, or exploring with a fusion of dishes, we savor the spice in our lives.

SCENTED GARDENS AND
AROMATIC HERBS

I come from a long line of gardeners and hope someday to live in a place where I can plant the offspring of my grandmother's peonies that, for now, live in my sister's garden and not my tropical Florida yard. The gardens of the American South, where I currently live, fascinate me as springtime brings the giant white flowers of magnolias, the creamy white blooms of gardenias, and the diffusive scent of jasmine on the warm and humid air. These white flowers make perfume for their pollinators, but the perfume also attracts us; we surround ourselves with such plants and tend them carefully. Herbs are garden staples that have brought freshness to homes when strewn on the floor or tended in simple bouquets. During times of sickness and plague the familiar scents of lavender and rosemary were fragile protections for the poor against the miasma of death and illness sweeping through the cities of Europe. Wealth and rare plants, on the other hand, make for grand gardens and rich displays—providing a place to retreat, "smell the flowers," and exclude the outside world. Gardens imply flowers, and flowers must have their pollinators, whether the moths that prefer a white flower's fragrance or the bees of the Mediterranean that flit among lavender and rosemary.

A scientist and an anthropologist wrote about the evolution of flowering plants. The scientist expressed frustration with the "abominable mystery" of their rapid evolution, and the anthropologist wrote about the world-changing coevolution between insects and plants that gave us flowers. The scientist was Charles Darwin, who in a letter to his son-in-law and fellow scientist Joseph Hooker stated that "the rapid development as far as we can judge of all the higher plants within recent geological times is an abominable mystery." Seventy-eight years later anthropologist Loren Eiseley published a beautiful essay called "How Flowers Changed the World" and ended with this sentence: "The weight of a petal has changed the face of the world and made it ours." We smile when we see a flower, be it a humble and happy daisy or a tall iris with its elegant shape and rich color. It is as if we somehow know we would not be here without flowers. Our relationship is deep and long, and it begins before we were humans: about 125 million years ago, when the first tiny flower may have poked its head above shallow waters or out from under dense ferns in a humid tropical forest. Up to this time land plants were pollinated by wind or water, using either spores as in ferns or pollen grains that could be picked up by the wind and, by a tiny chance, land on the sticky female cone of a conifer or gingko and pollinate it. Insects were present, including beetles and flies that perhaps found shelter on plants, ate some pollen or sticky secretions if they happened to find some, or possibly just hung out and chewed on stuff. If these insects assisted the plant in pollination, it was mostly by accident.[1]

The movement of earliest plants from sea to land began in or just before the Devonian period some 416 million years ago. These were primitive plants with early stems and leaves growing to perhaps a few feet tall. Next came conifers, ginkgoes, cycads, and ferns, classified as gymnosperms, or those with naked seeds. They grew tall and lush to make up the world's first forests—forests that carried out the important job of converting carbon dioxide to oxygen and supporting insects of enormous size. These primeval forests are the source of the massive carbon deposits that we mine as coal and petroleum. As gymnosperms proliferated, they provided food for insects and a growing diversity of land animals, including dinosaurs. It was after the Jurassic period, over 200 million years later, perhaps some time in the Creta-ceous, that plants began the transition to a form and habit that would attract and reward pollinators as well as protect their seeds. It was a time of increasing diversity of animals, including some of the first mammals and birds, along with many insect groups. Butterflies, ants, grasshoppers, and the first eusocial bee all appeared in the fossil record during the Cretaceous period, as did the first flowering plants, or angiosperms—plants that protected their seeds with covers called carpels—*angio* meaning vessel and *sperm* for seed. Along with these new seed types, a primitive flower arose, perhaps from the end of a green stem as a whorl of leaves that became specialized to protect the reproductive parts of the plant. Together, the evolution of larger, more colorful, and diverse flowers with carpel-bearing seeds allowed for im-mense success in the angiosperms.

Flowering plants multiplied and diversified with quick (in geo-logical time) exuberance, leading to Darwin's often quoted frustration with the abominable mystery of angiosperm divergence. Darwin liked to believe in the orderly and measured pace of evolution and mostly found evidence for it in his studies, insisting that "nature does not make a leap." But flowering plants did not follow this expected stately trajectory as they leaped and spread their beauty around the world. Darwin finally settled on two theories: first, a secret steady and slow-paced evolution of flowering plants according to his theories but in a secret land, perhaps an island or a lost continent that they finally es-caped from; or second, the potentially synergistic relationship between pollinating insects and flowers that propelled evolution not only of flowers but also of their pollinating insects. Science has seen more proof of the second theory than the first. Once the first flowering plant appeared it was not long, geologically speaking, before flowers became showy, displaying various colors and offering pollen as a reward. This meant that plants had to produce enough pollen and other rewards to ensure adequate supplies for both reproduction and attraction. As flowers grew larger, plants could pack more "stuff" into the blooms: like nectar-producing tissues and anthers with pollen. Nectar, pollen, and fragrance acted to bring in flying pollinators, and these partnerships between insect and flower resulted in an evolution of flowers that rapidly altered the world, at least in geological terms, and changed it completely because of the small, protected seed and the pollinators that visited flowers.

At the end, what flowering plants achieved was the ability to use a diverse set of reproductive strategies from simple wind pollination as seen in grasses to the intimate and specialized relationships such as that between some orchids and their pollinators. Small annuals could grow quickly and colonize new territory faster than some of the more ancient lineages, taking advantage of the wide landscape. Seeds were now protected by a covering and provided with sugars and starches to give seedlings a good start in the world and attract seed dispersers. Some plants developed lengthened pollen-accepting styles in the flowers to reduce the possibility of self-fertilization and increase potential for outcrossing. Nectar-producing bodies called nectaries appeared and functioned both to attract pollinators and to distract other insects from eating important plant parts. The shape of the flower diversified to become everything from a bowl-shaped welcome center for generalist pollinators to specific shapes that manipulate the type of pollinator and guide it to reproductive organs. Insects responded, and the world saw more flying pollinators, including bees, flies, flitting butterflies, and long-tongued moths.[2]

Flowers now make an array of fragrant volatiles to attract pollinators, leaves and flowers release compounds to protect against chewing and disease, and plants signal one another using those same fragrant volatiles. All organs of the plant from root to seed to flower may contain and release volatiles, but flowers are the masters of amount and diversity. As we will see, they may change their fragrance depending on environment, pollinator availability, and age of the flower. Scent may vary depending on the floral structure—petals, sex organs, nectar, and calyxes—or even in different parts of a petal. Some flowers are generalists and have a simple shape and a friendly blend of fragrances, but others manipulate flower shape, type of scent, timing of fragrance, and location of scent to pinpoint a specific type of pollinator and gain the advantage of increased reproduction and outcrossing. A plant may make and hold in their tissue constitutive volatile organic compounds at a certain level that generally defend against disease and herbivores, but they can also produce or increase on demand the same or specialized constituents to protect against a specific threat. Today there are very few habitats that flowering plants do not occupy, and angiosperms comprise about 90 percent of all land plant species.[3]

But when flowers appeared, we humans were nowhere even remotely on the horizon, so what does the evolution of flowers have to do with us? It was the nutritious seeds that fed the small mammals of the time, allowing them to prosper in the face of global change that included extinction of the dinosaurs. Up to that point mammals were little and innocuous, eating insects and hiding during the day, but now they could come out from the forest shadows and develop the chewing teeth that allowed them to eat tough flowering plants and seeds. Getting rid of scary dinosaurs probably was not enough to allow mammals to evolve and expand: the new plant foods helped mammals to diversify, but it was also insect pollinators that arose and expanded to take advantage of the new food sources. Agile mammalian insectivores were primed to come out of hiding and take advantage of these little packets of flying food. A diversity of mammals occurred, and eventually primates appeared to give rise to early humans, who walked out of Africa about two million years ago.

We value flowers simply for what they add to life. And to death. Beauty without commerce and comfort without gain—flowers for graves are an ancient practice even when the act of gathering them took away from the critical activities of sustaining life. And yet, as much as 13,700–11,700 years ago in the Raqefet Cave on Mount Carmel in Israel four Natufian people were laid to rest in flower-lined graves. Sage, mints, and snapdragons were carried in to line the graves, and they have left delicate imprints of flower and stem under and around the bones.[4] Floral pollen has been found alongside the bones of Neanderthals in the Shanidar Cave of northern Iraq and was thought to have been evidence of ritual burial. But gerbil-like mammals live in caves like this and will collect flowers, bringing them in and leaving pollen behind, leading scientists to question the origin of the pollen. Renewed research in the Shanidar Cave seems to be gathering more evidence of ritual burials in these caves, with or without flowers. Our appreciation of flowers continues through time, not only to give us peace in times of sorrow but to celebrate just about any occasion—and it has been commercialized in a big way. We spend our hard-earned money to buy seeds and plants, we spend our weekends tending to our flower beds, and we hope to have a few blooms to grace our dinner table. Or maybe we just gather them on our nature hikes to keep in a vase until they sag and wilt or to give to a loved one as we appreciate simple acts and the fragile weight of a petal.

Coyote tobacco (Nicotiana attenuata) *flower with sphinx moth*
(Manduca sexta)

6

Gardens

We do not know when the first plants were sowed or transplanted with a purpose or who made the first garden, but we do know that people have had gardens for thousands of years. Common sense and history record them as being useful for growing food or medicine and for providing religious symbols and quiet retreats. At some point humans began to tame, cultivate, and breed flowers in our gardens. But what defines a garden, and where might they have originated? We have seen how easy it might have been for someone living in the Western Ghats of India to plant a few black pepper vines near their home or for an early Mesoamerican to move a vanilla vine or two from the rain forest to a nearby tree that was perhaps more accessible. As early as 4,500 years ago dwellers in the eastern Amazon were selecting edible forest species, selectively clearing forest to cultivate crops, and using limited fire to encourage the growth of these food sources within their forests.[1] Even today edible plant species are abundant in the eastern Amazon forests. Forest glades, peaceful hillsides, or flower-strewn meadows may have inspired early flower gardens; perhaps a convenient fallen log was moved to a vista so locals could sit and enjoy the quiet view, and a few bulbs were moved to a nearby sunny spot. But when did a garden become a garden? Maybe it does not really matter, and we can accept that "garden" is in the eye of the beholder; and as both beholder and gardener I have a loose definition of what *garden* means. I am thinking of the accidental wildflower lawn of a neighbor where small weeds grew and flowered among tough grass in between infrequent mowing and lackadaisical care. We chatted about how much I liked his yard and enjoyed watching butterflies and bees congregate over the weedy flowers. At our house, on the other hand, we did away with any pretense of a lawn and tended our woodsy garden that had flowering trees, native shrubs, and passion vines to feed hungry caterpillars. Zebra longwing caterpillars (*Heliconius charithonia*) would eat the vines bare and make their alien-looking pupae to create more graceful butterflies that would roost at night under the palm trees of our south Florida yard. Fluttering wings would appear just as the sun was setting as the half dozen or more butterflies found their perfect twig to hang from. There was the occasional battle conducted in slow motion, with much flapping of fragile wings, until all was settled for the night.

Perhaps the first gardens were unconstructed and individual— rather than getting the exact same plants as everyone else from the local big box store, homes and families and villages would sequester and protect a small area with a primitive fence of rocks or wattles or even some slender willows planted close together. They chose edible or medicinal plants that were useful and that would easily transport from nearby woods and fields, consulting with neighbors and sharing their knowledge and harvests. Pretty flowers were almost certainly grown as well, perhaps simply for the joy and beauty they provided and for sharing with friends and neighbors. Home gardens are intimate spaces created for a household and often provide supplemental food as well as medicine. These are one of the oldest and most persistent forms of cultivation and nearly always come under the supervision and creation of women in the household. Even today such activities support local plant types such as heritage plants that inspire regional foods and may also represent unique genetic types. Gardens are living things, ephemeral, they require tending, and they have long been found in dooryards of the humblest homes, in monasteries, and in palaces. A garden's purpose is reflected in its structure, from the utilitarian kitchen garden to gardens created for pleasure and refuge and to elaborate walled gardens that demonstrate power and wealth. Whichever shape they take, we can see some commonalities in gardens large and small. They are often cordoned off: think of the wattle fencing during the Middle Ages that protected aromatic roses and held in the scent, or the grand fences around palace gardens that kept out the riffraff, or perhaps the fragile wire fencing that supports string beans on the perimeter of an urban community garden.

Beneath every garden there is dirt. Pliny the Elder in his work *Natural History* reminds us that "indeed, the genuine rose, for the most part, is indebted for its qualities to the nature of the soil." We appreciate and sometimes tend the soil in our gardens but otherwise probably do not think about it much. There is, however, a place in India called Kannauj where they consider the scent of the soil to be a perfume comparable to the finest rose or jasmine: they call it mitti attar. During the dry season, the earth in this place soaks up tiny scented molecules secreted by plants. Dryness draws in the molecules, layer upon layer, and heat bakes in the fragrance. Monsoons come, as they will, riding on a wave of moisture that liberates the scent of the baked-in fragrances and releases the perfume called petrichor. But the distillers of Kannauj do not wait for the monsoons; they want to capture that scent before the rains come. As generations of families have done before them, they gather the soil and dry it into disks that are taken to traditional deg bapka stills. These stills are ancient and made of copper and bamboo placed over brick fireplaces. Essential oils from the soil disks are distilled, carried through a bamboo pipe into bapka receivers containing sandalwood oil and cooled by water in a trough a level below the copper degs. Excess water from the distillation is drained from the bottom of the receiver (because essential oils float), leaving the fragrant, soil-laced sandalwood oil behind. This oil is transferred to leather containers called kuppis, where it is sealed in and left for any remaining water to evaporate and the rich, loamy scent of the mitti attar to bloom.[2]

Egyptians and Persians, both desert peoples, valued walled gardens for their sense of retreat and beauty. Herbs in beds, palms and pomegranate trees for shade, and ponds with iconic lotus flowers are depicted on the walls of Egyptian tombs. Colorful poppies and cornflowers provided reds and blues interspersed with yellow fruits of the flowering mandrake, while papyrus completed the bouquet. Persian gardens were enclosed within sheltering walls where the sound of water, trees full of fruit, scented and flowering plants, and pockets of shade could all be found. An overall garden plan had water channels meeting at right angles to divide the enclosed space into rectangles and wide basins, flowerbeds that were sunken for ease of watering, and courtyards with rows of cypress or poplar trees. Orchards of almonds, apricots, plums, pears, pomegranates, and wild cherries also bloomed in the gardens. Flowers such as roses, jasmine, anemones, tulips, irises, and violets were cultivated, and aromatic herbs grew among the citrus trees. Gardens and agriculture in arid zones of Persia were supported by *qanat* systems of irrigation, underground tunnels that tapped into alluvial aquifers to transport water, now recognized by UNESCO as world heritage sites. The Ottomans had a rich garden culture and would grow and export tulips, hyacinths, and irises into Europe. Persian garden designs inspired the pattern for beautiful and intricate woolen carpets that were filled with stylized flowers and trees. Eventually more complex structures of pleasure and romance were built, such as the Taj Mahal, which honored Empress Mumtaz Mahal, favorite wife of the Mughal emperor Shah Jahan, and the tomb garden memorializing the Persian poet Hafez in the Musalla Gardens of Shiraz, Iran.[3]

Chinese gardens were an integral part of the arts of poetry, calligraphy, landscape painting, and gardening. They were often designed to unfold in the manner of a scroll for appreciation and were marked by unusual rocks and arrangements of stone. Orchid, bamboo, chrysanthemum, and flowering plum represented the four seasons and the attributes of grace, resilience, nobility, and endurance. The King of Flowers, tree peonies, with perfect blossoms and an ancient heritage going back as far as the third century AD, were cultivated, and lotuses grew their graceful flowers to rise above the muddy dirt and emerge from clear water.

In contrast, somewhere in the sixteenth and seventeenth centuries most Japanese gardens became focused on an austere Zen approach that complemented their tea ceremony. Flowers are more likely to be found in restrained indoor ikebana arrangements, and the garden's purpose is to set the mood for the ceremony. If there are stepping-stones, they are arranged to control the pace and direction as one moves toward the tea house. Rather than distract, stones and a single tree such as a willow or maple invite participants to pace themselves, pause, look up, and take a moment. But the Japanese also enjoy the ritual of viewing abundant displays of cherry blossoms that have, for centuries, been a national pastime celebrated by families and people from all over the world. In fact, during the ninth century, women of the court would wear kimonos in colors that matched the blooms. The modern tourist to Japan can book tours to various scenic cherry blossom localities and perhaps find matching clothing.

Red roses, gillyflowers, purple hyacinths, and marigolds flowered in a medieval European garden and might have been planted with feathery parsley or fragrant mint. Sage and lilies, iris and fennel, and rue and tansy added their pretty flowers, and walls would contain their fragrances on warm spring days. Fragrant herbs such as thyme and basil were mixed with straw to cover floors inside and add fragrance as they were walked upon. Pots of lavender and rosemary grew on windowsills, and a bit of mint could be used to add flavor to a stew. Or perhaps rue, sometimes called the herb of grace, added its nearly unpleasant (and sometimes toxic) sharpness and blue cheese taste to small dooryards both in its native Mediterranean but also as a transplant to the New World. The roots of the iris provided starch and a light scent for the laundry, and the yellow flowers of tansy were an important medicine to treat worms and digestive problems. For the wealthier, larger gardens were more organized—highly detailed woodcuts of the time generally show some sort of wattle fence and roses with perhaps a gardener off to one side leaning on a tool of some sort and a lady sitting among the roses. Gardens were often laid out in geometrical shapes with paths between each garden area as well as a grassy lawn, perhaps with a bit of raised turf or tuffet for sitting.

While Europeans may have had their own small plots with useful plants, it was the apothecary garden, often associated with monasteries, that provided a home for the various medicinal plants. Hildegard von Bingen, born in 1098, was a German abbess who documented medicinal herbs for the garden while also writing music and studying math—she was an early source of knowledge regarding European herbs. She died in 1179 after a prolific life of consulting with other monasteries, writing, composing music, and creating art in a wonderful synesthetic blend. In an apostolic letter in 2012 proclaiming her a doctor of the universal church, Pope Benedict XVI recognized the relation between scent and saintliness when he noted that Saint Hildegard "died in the odour of sanctity." Across the world and somewhat later in the Americas, Spaniard explorers discovered the floating gardens of the Aztecs that were planted around Mexico City sometime around the fourteenth century. The gardens were planted on raised beds in linear patterns along the edges and even within the lakes as floating or raised islands, and they produced aromatic herbs,

medicinal plants, a variety of vegetables, and flowers for the wealthy.[4]
Two written works dating from the time of the Spanish conquest, the
Florentine Codex and the Badianus Codex, describe local herbs and
plants used for medicine and are illustrated with beautiful and vibrant
drawings of the plants and their growing habits.

As the European world entered the Renaissance, a garden's purpose moved beyond the simple provision of food and medicinal herbs
to become a place of refuge and display. The wealthy could leave cities
where the plague raged and find order, peace, and beauty in flowers,
lawns, and running water. Fragrance was also an intrinsic part of
many designs because pleasant aromas were thought to be a force of
good to counter the fetid smells of disease. These gardens had a
plan—they had green lawns, statuary and fountains, walls, and maybe
even a tame rabbit or two. Unusual and fragrant plants such as tuberose and scented geraniums traveled from exotic locations to become
garden favorites. Sometime around the sixteenth century new plants
began coming into Europe from both the East and the Americas to
inspire gardeners. Explorers from the Americas sent potatoes, tomatoes, sunflowers, nasturtiums, and tobacco, while Turkey sent the gorgeous flowers of hyacinth and tulip.

By the seventeenth century formal gardens in France and Italy
were constructed with paths, knots, and patterns of herbs to walk on
and brush up against, perhaps violets, strawberry, wild thyme, and
water mint. Formal gardens as planned and enjoyed by royalty and
the wealthy, like most early gardens, were enclosed, but these walls
were to keep people out. Some were decorated in the current style,
which included grottos and ruins to evoke a sense of ancient Greeks
and Romans, as well as fountains, mazes, and shaped hedges that
made them predictable, formal, clean, and perfect for gathering with
friends and peers. Was it ostentation or grief that inspired the very
unusual Sacro Bosco (1552) near the Italian town of Bomarzo? Quite

different from the traditional Italian Renaissance garden, it was left mostly natural with wandering paths to take the visitor from one surprise to another. Sea monsters, off-kilter grottos, battling creatures, Cerberus, and giants made of stone fill the garden. There is a giant head with a wide-open mouth forming an enclosure where people could simultaneously eat while being eaten and view the nearby inscription that reads "All reason departs." The owner, Pier Francesco Orsini, duke of Bomarzo, had returned from war, where he saw his friend killed and he had been held prisoner; shortly after his release, his wife died. The garden fell into disrepair starting in the nineteenth century. In 1937, the artist Salvador Dalí visited the overgrown ruins and fell in love, making a short film there and using the sculptures as inspiration. Since restored and in private hands, the park is now a popular tourist attraction.[5]

Louis XIV, the Sun King, had the gardens at Versailles designed in a very traditional and stylized manner, with a main allée over a mile in length that separated water and woodland features. But his landscape architect also inserted surprising little byways and shady groves to contrast with stylized lanes and sunlit spaces. Within the next century, European gardens moved from strictly stylized and separate from the world to more outward-looking and natural designs. Fences were minimized, and structures like moats and recessed walls (called ha-has) were devised to keep out cattle (not invaders) and allow a more seamless transition with the landscape. Gertrude Jekyll was a prolific writer, painter, and horticulturist who created over four hundred gardens in the nineteenth century. Her partnership with the English architect Edwin Lutyens was highly influential in the Arts and Crafts movement. Jekyll designed for color, being influenced by the artist J. M. W. Turner, and would designate cold colors such as white and blue to contrast with hot colors such as red and orange and was known for her herbaceous borders. In case you were wondering, her brother was friends with Robert Louis Stevenson, who used the family name in his novel *Dr. Jekyll and Mr. Hyde.*[6]

Urban gardens have been with us for a long, long time, and any distinction between urban garden and, well, garden is going to be fuzzy and perhaps somewhat arbitrary, but there are a few that bear mentioning. Aztecs built the famous floating gardens in Tenochtitlan to provide food security for the city but also to grow flowers and herbs. Japan has a long history of growing and cultivating the sakura, or cherry blossom, and celebrating the blooms each spring in cities throughout the country. Multitudes of locals and tourists stroll among the flowers, bringing picnic lunches and socializing. City inhabitants are rediscovering the power of planting gardens in food deserts to provide food and flowers for locals, and school gardens are helping children learn to garden and reap what they sow. Botanical gardens are some of my favorite places to visit when I travel and were created in cities and associated with universities as teaching tools, the first in Padua, Italy, in 1545 followed by Leiden, Leipzig, Heidelberg, and Oxford up through 1621. Singapore has upped the ante on the urban garden at the Jewel Changi Airport, where a series of gardens has been created for the traveler that includes a walk through a forest, an abundance of flowers, and waterfalls.

Plant exchange was a by-product of trade along the routes between East and West and the campaigns of conquerors, including Alexander the Great, Islamic warriors, Genghis Khan, and the Crusaders. But the plant hunter is a somewhat different beast, sent out specifically in search of a diversity of plants: European plant hunters of the eighteenth and nineteenth centuries sought interesting and different—but also profitable—plants for their sponsors. They brought back scented geraniums from southern Africa, large conifers from the western United States, and rhododendrons and azaleas from Japan. But for exploration and travel specifically to bring back unusual and aromatic plants, we must begin with Queen Hatshepsut, pharaoh of Egypt, who might be called the first of the plant hunters for organizing the legendary expedition to the Land of Punt in search of aromatic plants. Her ships returned with incense trees, likely frankincense and myrrh, as well as gold and other trade goods. Another Egyptian ruler, Thutmose III, collected a variety of plants from his expeditions throughout Asia and featured them in the decorations of his sacred and symbolic botanical garden room at the temple complex of Karnak.

Move forward nearly three thousand years, and plant hunters were bringing plants from around the world to the royal gardens at Kew and to European nursery owners for sale to the wealthy. In the early seventeenth century, the naturalist John Tradescant worked for the first earl of Salisbury and, after beginning in the Netherlands, Belgium, and France in a hunt for tulips, roses, and fruit trees, he became a professional plant hunter. In addition to creating Tradescant's Ark of Curiosities, a collection of natural history objects, he was His Majesty's Keeper of the Gardens, Vines, and Silkworms at a royal palace in Surrey. Sir Hans Sloane was a botanist who studied at both the Apothecaries' Hall in the City of London and the Chelsea Physic Garden. He traveled to the West Indies, acting as physician, where he tasted the local chocolate drink and found it unpalatable. As he tried to improve the taste, he mixed the cocoa with milk and sugar and created Sir Hans Sloane's Milk Drinking Chocolate (or what we would call hot chocolate), a recipe eventually bought by Cadbury, the chocolate company. After decades of travel around the world, collecting some 71,000 objects along the way, he bequeathed his collection and notes to the nation of Britain in return for £20,000 to his heirs. Parliament accepted his terms in 1753, jumpstarting the establishment of the British Museum, which opened its doors in 1759. Sir Joseph Hooker was a prodigious traveler, visiting Africa, Australia, New Zealand, South

America, and India in the mid-1800s. The rhododendrons he gathered from the Indian state of Sikkim came at great cost as he climbed mountains in severe spring weather at high altitudes, and descendants can still be seen at the Royal Botanic Gardens, Kew, in London, in the Rhododendron Dell, but be sure to go in spring for the best blooms.[7]

Sir Joseph Banks traveled with Captain James Cook between 1768 and 1771 and brought back some 3,600 dried plants, 1,400 of which were new to Western science, but was denied a second voyage due to his excessive demands for accommodations and personnel which included two French horn players. Banks became unofficial director of Kew Gardens and was responsible for sending plant explorers around the world. One of those missions was that of Captain Bligh for the purpose of collecting breadfruit to grow in the West Indies, a voyage that resulted in the infamous mutiny on HMS *Bounty*. Bligh was sent back to Tahiti after being acquitted of wrongdoing and brought back 349 species of plants. In the Western Hemisphere, explorers to the Amazon discovered thousands of plants, including the cinchona tree—a source for quinine, which is used to treat malaria. In the newly formed United States, Thomas Jefferson was an avid gardener and interested in both horticulture and agriculture. In his allées, he had his favorites or pet trees, which I totally understand. Jefferson originated and funded the expedition of Meriwether Lewis and William Clark in 1804 to learn more about the geography and botany of this new country. They returned with about 182 plant species, over half of which were new to science, including the prairie wild rose (*Rosa arkansana*), and they described native tobacco (*Nicotiana quadrivalvis*), plus a variety of other plants and animals, in their detailed journals written in the roughest of conditions. Decades earlier, among a growing abundance of plants new to Western scientists, Carolus Linnaeus took on the responsibility of bringing order using sex organs as the basis for a precise binomial nomenclature, consisting of genus and species, that uniquely identified each group of plants within an organized hierarchy of relationships. His classification scheme expanded to all known living beings that are now identified as to their place in the living world.

An important and yet seemingly small invention provided the way for explorers to maintain and deliver live plants through long and challenging voyages. Wardian cases originated around 1833 and were basically small, portable greenhouses that allowed for transport and maintenance of delicate plants. Many explorers of the day were tasked with bringing back interesting plants from far-off places, and for many years they were restricted to seeds and cuttings from tough plants. Or they may have done as David Fairchild (who lent his name to the Fairchild Tropical Botanic Garden south of Miami) did and stick their cuttings in potatoes or moss (in his case, citron as he escaped on a mule from a Corsican grove) to keep them alive.[8] A certain Dr. Ward was a fern enthusiast and had much difficulty growing ferns in his rockery in the cold and smoky environs of London until he took a closer look at a failed experiment. He had tried to hatch a sphinx moth caterpillar in a bottle that developed mold at the bottom. Instead of the moth, he saw a fern growing out of the mold and decided that the moist environment with controlled heat and light was what the fern needed to survive. Expanding on that concept, he constructed small glasshouses for his own use that became the model for Wardian cases used by botanists to transport fragile botanical specimens through harsh oceanic conditions for months at a time back to England and Continental Europe.

Modern plant hunters may fall into the category of scientists searching for botanical medicines or unusual species, but the story I quite enjoy is of the Texas Rose Rustlers. A small group of Texans, around the 1980s, began spotting neglected-looking roses that bloomed and sprawled in the heat of local summers. Taking clippings where possible, they began to revive these antiques originally brought to the South from Europe to be planted around stately houses. These beautifully scented roses were formerly nurtured in local gardens until a newly fashionable hybrid rose made its appearance and the "old roses" became relegated to neglected areas, where they became pass-along plants growing in humble farmhouses and graveyards. Texas Rose Rustlers were passionate about their cause and found that the antique roses thrived in their gardens without the watering and pesticides that hybrid roses required. Among the requirements for Rose Rustlers were sharp shears, plenty of insect repellent, an honest face, the ability to say "Don't shoot!" in several languages, plastic bags, and a Sense of Mission. But they were also required to pledge not to trespass or remove plants without permission.[9]

Rosemary (Salvia rosmarinus) *leaves and flowers*

7

Fragrant Flowers and Aromatic Herbs

People have manifested their love of flowers through art throughout history, in the drawings of crocuses in Mediterranean Bronze Age art and in the highly stylized rugs and mosaics of Persian gardens. We have been blessed with Monet's lilies and Van Gogh's sunflowers, vibrant drawings of important local flora by Aztecs, and detailed works by Rumphius and many other talented botanical artists through time. But some of my favorites are the woodcuts of the Middle Ages showing humble wattle-enclosed gardens with roses and their keepers. Perhaps the act of gardening itself reflects art through the flowers, colors, shapes, and sizes we choose to display. Alice Walker in her book *In Search of Our Mothers' Gardens* writes of the need for beauty and self-expression that was illustrated by her mother's gardening: "Guided by my heritage of a love of beauty and a respect for strength—in search of my mother's garden, I found my own." Her mother toiled day after day in fields not her own and yet came home to plant dazzling arrays of profusely blooming plants, her expression of art. Gardens of share-croppers and enslaved workers of the American South were the product of necessity, passion, and muscle as they bent their backs to plant and nurture vegetables and flowers for their homes. Gathering plants from the woods or trading with neighbors, they composted hay and manure, tilled the soil, tended seedlings, and harvested in between labors in gardens and fields of others. Nearby woods and waterways provided places to meet, seagrass and oak for baskets, medicines, and even visual inspiration for art like the patterns of practical quilts. They learned from each other and from experience, creating truck gardens and flower patches to grow iconic southern vegetables such as okra, collards, watermelon, and sweet potatoes for necessity and flowers such as petunias, daylilies, cannas, roses, azaleas, camellias, and wild honeysuckle for beauty.[1]

Most of us appreciate the visual beauty of a garden and may include a fragrant or a fuzzy plant for the other senses of smell and touch. For the visually impaired, as well as others with special needs, sensory gardens allow visitors to enjoy and experience flowers, herbs, and vegetables in their own way. Such gardens may focus on scent but also highlight sound, touch, and taste: they are often designed to be fully accessible to wheelchairs and easily navigated by the blind. Visitors may brush up against fragrant scented geraniums, rub a zesty mint plant, or walk on creeping thyme alternating with smooth river rocks. Night gardens are another type where vision recedes and fragrance and sound take over. Imagine dusk in the garden: perhaps it is walled, and the air is still and quiet. You begin to walk the paths and stop, backtrack a little, and sniff. There it is! The very definition of floral in a quick waft of sweetness. You stand for a moment and look around. Within the deep green shadows, you see the last fading light reflected from a scattering of white flowers, white sand, and silvery foliage that rustles with the breeze. You sit on the bench conveniently provided for the appreciation of the blooms and enjoy the simple experience of fragrance, solitude, and quiet. But if you listen closely you can hear the whir of wings and, if you watch, you can see what looks like a hummingbird hovering over the white flowers. A closer look reveals that it is not a hummingbird but a moth: a sphinx moth, which is the nocturnal insect equivalent of a hummingbird. The moth has also followed the scent plume of the white flowers and is sipping nectar with its long tongue while it hovers in front of the bloom. As it sips the nectar, its hairy body gathers pollen from the flower that it will deposit when it drinks from the next bloom. As a perfumer chooses the materials and ratios to create a perfect blend, so do white flowers make and release their own "odor code" into the night air to lure winged pollinators. White flowers almost always are night blooming and include such iconic plants as gardenias and magnolias, evening-scented stock and fragrant tobacco, brugmansias and night-blooming jasmine, tuberose and poet's jasmine.

Gardenia jasminoides is often simply called gardenia but also Cape jasmine due to a mistaken identity and place of origin—the plant was originally thought to be a jasmine relative and to be from the Cape Peninsula of southern Africa instead of its native China. Gardenias are widely cultivated and appreciated for their large flowers and gorgeous scent. The complex and nuanced fragrance of gardenia is, to my mind, at its best in the old walled gardens of the American South, where humid summer air gathers the scent in all its richness. Once upon a time in an old walled garden in Charleston, South Carolina, I wandered around a corner to find a fountain splashing in the corner, an old iron bench, and a huge gardenia bush spreading its fragrance in the air. With newly opened buds of pure white and older flowers of a yellowed ivory color, the fragrance was just right as it was borne on the still air: I could move closer and bury my nose in a flower or retire to the bench and enjoy a more diffuse and subtle aroma. I can find a list of the constituents of gardenia in the literature and the list may vary according to method or flower type, but words on the page cannot capture the dance of molecules that the flower throws into the air as it blooms. Some make their presence known with a sharp and fresh greenness over buttery and milky tones and the occasional indolic mushroomy scent that appears and disappears depending on the time of day. Some may incorporate sweetly floral linalool as well as terpenes for a spicy, woody, and green fragrance; the minty smell of methyl salicylate; and a bit of indole. Since they are pollinated by moths, the fragrance is strongest at night but is also present in early morning and late afternoon. On a cloudy day in Charleston the flowers can fill a small walled garden with fragrance. Gardenias also came in handy in the smoky and boozy atmosphere of American jazz clubs where Billie Holiday was known for wearing one in her hair.

Before we delve into the fascinating and complex world of tobacco, moths, wolf spiders, and hummingbirds, this seems a good place to take a moment to talk about the tools that plants use to carry out life's necessary functions. One of the primary ones is reproduction, which plants must accomplish from their rooted location by reaching out to the intermediaries that move male gametes to female gametes. Some plants are self-fertilizing, like black pepper, so weather and wind can do much of the work: flowers on these plants are small, and investment in nectar is minimal. Some plants reach out a little farther but still use wind or water to carry pollen to others of the same species: out of many grains released to the wind, a few will fertilize another plant and perhaps bring together new combinations of genes to increase fitness. These flowers are also inconspicuous, but they produce prodigious pollen noticeable to anyone that suffers from hay fever or a pollen allergy. Others invest in floral resources: they make rewards such as nectar that can be siphoned out by long tongues of moths and protein-rich, sticky pollen that is removed by bee or bat for consumption and will stick to a hairy body to be carried to the next flower.

Sometimes the relationship between plant and pollinator becomes so specialized that we can predict which pollinator is attracted to, and best suited for, each flower.[2] Pollinator syndromes have been described since the 1960s and detail a coordination of flower color, flower shape, nectar volume, nectar concentration, and fragrance blends that attract specific pollinators. Scientists now advocate for less than complete fidelity to these categorizations and use them more as guides—but any gardeners and hikers can enjoy their outdoor experience as they try to match pollinator with flower.

I described microcantharophily, or beetle pollination, in connection with the nutmeg family, but if you take away the *micro-* part of the word, cantharophilous describes magnolias and lotuses with their large bowl-shaped flowers, light colors, and a strong, often fruity fragrance. Larger and messier beetles seem to be attracted to these flowers. Bees are attracted to flowers in the pink-purple-blue range but also white or yellow, and the flowers often have nectar guides—patterns in contrasting colors that may also reflect ultraviolet light. Melittophilous, or bee-loving, flowers have a sweet fragrance and provide a moderate amount of nectar to their visitors. Butterflies forage on yellow, orange, mauve, or red flowers in the daytime and may respond to nectar guides and a light fragrance. Often psychophilous, or butterfly, flowers have massed blooms with long tubes that fit the coiled proboscis of the butterfly as it probes for the hidden dilute nectar. When evening falls, the moths come out, and they can be separated into two foraging types: settling moths and hovering moths. Both types pollinate highly scented flowers that bloom between dusk and dawn on such plants as jasmine, tobacco, and gardenia. Night-blooming flowers are lightly colored in white, cream, or pale green, and they are highly scented with a sweet fragrance and moderately concentrated nectar found at the end of a tube. Sphingophilous flowers send their perfume out to sphinx or hawk moths with long proboscises that are delicately inserted into long nectar tubes while the moth hovers in front of the flower. Phalaenophilous flowers have tubes of moderate length and are pollinated by settling moths that land on the petals and have tongues that match the shorter length of the nectar tube. Bats respond to the light colors of night-blooming chiropterophilous (bat-loving) plants such as many desert cacti, agaves, and the legendary baobab tree, finding flowers by following the fruity-fermented or musty fragrance of some or by echolocation on the large blooms. Ornithophily describes pollination by birds like hummingbirds but also honeyguides, sunbirds, and others that prefer red and orange flowers with a short nectar tube full of sweetness that fits their bill but do not necessarily have a fragrance. Red is a definite hummingbird attractor as I learned while doing fieldwork in the deserts of southern Nevada. My favorite shirt was a white t-shirt with a large red hibiscus on the front that invariably drew the attention of black-chinned hummingbirds (*Archilochus alexandri*), who hovered in front as they evaluated the nectar potential of the fake flower.

Pollination by flies is called myophily and may describe flowers with white or green color and a mild fragrance. Not all flowers are sweet and pretty, and these are the ones that practice sapromyophily, or carrion-fly pollination. These flowers are often more dramatic, come in shades of dark red or brown-purple, and smell like rotting meat or feces. This stinky method is used by the world's largest flowers, in the genus *Rafflesia,* which grow underground in Philippine and Indonesian forests and are parasitic on tree roots. The flower appears at ground level as a huge reddish-brown inflorescence. Seeking rotting flesh, carrion pollinators like blowflies enter grooves in the male flower, pick up pollen, and then visit other flowers, perhaps landing on a female flower and fertilizing it.[3] Of course, there is also the very stinky titan arum, aka the corpse flower (*Amorphophallus titanum*), with a huge and foul-smelling inflorescence containing many small flowers that blooms rarely and always attracts large numbers of visitors to revel in the size and stink at botanical gardens where they reside.

Without tooth and claw, plants find a different way to protect themselves: thorny stems and prickles work for some plants, but often the more delicate leaves and flowers require chemical assistance. This means that many plants have become chemical factories, producing a variety of volatile aromatics to assist in reproduction and protection. Over 1,700 volatile organic compounds, or VOCs, have been identified in over 990 plant species (these are the ones that scientists have taken the time to analyze), and they are tools that accomplish some of life's necessary activities. Whether emitted into the air or accumulated in the tissue of a plant, they are defensive or protective, sometimes attractive, certainly communicative, and they must fulfill those roles with regard to pollinating, chewing, and boring insects; mammals and other grazers; microbes; and neighboring plants. Time of day, season, and even altitude may influence the presence and release of VOCs, and it is important to understand which tissue is producing the molecule to determine its purpose.

These volatiles are also referred to as plant secondary metabolites because they are not actually necessary for such life functions as growth, respiration, and reproduction. They may be produced and remain in plant tissues to protect against damage from herbivores, harsh environmental conditions, and/or disease, as we saw with sandalwood, agarwood, and resin-producing trees that retain terpenes and sesquiterpenes. Leaves and stems may contain protective compounds within their tissues but are also able to release molecules that are activated if they are damaged. These are called green leaf volatiles and will be familiar to anyone who has mowed their lawn or trimmed their hedges—they provide that sharp green smell that is released by damage from chewing insect, grazing mammal, mower, or clippers. Seeds like spices are protected by pungent and aromatic compounds, and seeds in fruits are enclosed in fragrant and tasty packages so they are eaten and dispersed along with a bit of fertilizer in an animal's feces. Then there are the chemicals that so many plants produce in their flowers that are specifically for attraction and are released into the air. These are often terpenes but, as we saw with gardenias, are generally a mix of various volatiles that are blended within the flower or its parts. The formula of the scent, the timing of its release, and the location within the flower are all tools that each plant has evolved to attract the pollinator that serves it best. Because production of these molecules requires an investment on the part of the plant, there are also controls within many flowers that can fine-tune formula, scent, and timing and even curtail the production of scent if a pollinator does not respond to fragrance. Hummingbirds, for example, do not have a sense of smell and respond to color and shape, meaning that a flower would be wasting precious resources if it attempted to attract a hummingbird using fragrance.[4]

There is a delicate dance between the genes that code for production of these molecules and it includes DNA, enzymes, and the assembly line that adjusts precursor molecules present in the plants and turns them into pollinator attractants. The dance may involve time of day, elevation, microclimate, weather, geographic variation, and type of pollinator. These genes are answerable to selection, usually from the actions of pollinators, and have evolved to give us the fragrance of flowers. The effect of scented compounds may be further modulated through the release of different constituents from various parts of the flower. Volatiles from petals generally travel out into the air to grab the attention of pollinators in search of nectar or pollen. Once a pollinator approaches the flower, it may find colorful nectar guides on the petals that lead the way to nectar or the pollen, a particular shape to encourage or discourage, or a flower with a position on the stem that may or may not encourage exploration. Nectar or pollen sometimes has a distinct fragrance to which a bright yellow color may be added to pull in pollen eaters. Once a flower is pollinated, its fragrance may alter to inform pollinators that their job has been accomplished and they should move on to the next plant or to a different open, unpollinated flower on the same plant.

In the ongoing quest to identify the elusive aromatic constituents of flowers, the development of headspace analysis allows scientists and perfumers to take a tiny flower, or even a part of a flower, and determine the composition of its fragrance in the field. Flowers and even plants often emit hundreds of different fragrant molecules, and the human nose quickly reaches its limits. Scientists will take a tiny container lined with a fragrance-capturing surface and carefully insert a single flower or a scented leaf, even one still attached to the plant. The captured scent can then be eluted off and analyzed using a gas chromatograph (GC), which separates each molecule based on chemical properties and passes them to a mass spectrometer (MS), which identifies the molecules in a process abbreviated as GC/MS, spitting out a long list of constituents. Headspace analysis has allowed scientists to plunge into the fragrance chemistry of flowers and plants so they can identify important individual scent components. Whether scientists are looking for the blend that attracts a moth or the chemical that signals damage, the information puts on paper the complex and mysterious fragrances emitted by the plant. Do this for tobacco flowers, and the result is a long list of fragrant chemicals, many of which are obscure, but the general agreement is that it has a "white flower" perfume—floral, sweet, and rich. Which, of course, anyone who has stuck a nose up to a tobacco flower will know.

Soil, insects, sun, fragrance, and rain do the work in the garden and give us flowers—the crowning glory of our gardens—while we do our bit to tend and tame the plants. Flowers, the sex organs of plants, come in all shapes, sizes, and colors: not surprisingly, the showy hibiscus and the fragrant tobacco have two quite different ways of attracting their different pollinators. Tobacco is all about the volatiles and is an expert manipulator of both pollinators and predators with a variety of fragrant tools at its disposal. Considered a white flower, tobacco is also a plant of contradictions. Most of us think of the plant as the source of the addictive and harmful substance nicotine, but the genus *Nicotiana* contains about sixty species, and only a few are used for smoking tobacco. Growers of fragrant gardens are familiar with flowering tobacco, a nocturnal bloomer and moth favorite. All have star-shaped tubular flowers, tend to be tall and floppy, and have long been used by gardeners to add a graceful appearance and beautiful fragrance to a garden as they blow their perfume out into the night air to reach the feathery antennae of moths. *Nicotiana* species are found primarily in South and North America but also in Australia and on some South Pacific Islands. They are members of the Solanaceae, or nightshade family, and species tend to be found in warmer habitats. Outside the tropics most tobacco species are annuals, dying off each winter to reseed or sprout from the roots in the spring. Where they grow wild, they can be found in a variety of habitats, including weedy and disturbed areas. Attracted by the white tube-shaped flowers and the floral fragrance, moths, bees, and hummingbirds pollinate the plant, but it is a particular kind of moth that scientists have studied to understand the relationship between flower and plant, pollinator and predator, and the fragrance that ties them together.

As scientists have studied the relationship between white flowers and moths, they have described a few characteristics of these moth-pollinated blooms. White flowers are generally light in color with a dissected shape to the petals for better visibility against the darker leaves; behind the petals is a tubular neck that fits the long tongue of the moth (or does the tongue of the moth fit the tubular neck of the flower?); they have dilute nectar that encourages the moth to visit a succession of flowers rather than filling up at its first visit; and their sweet, highly floral, often complex perfume causes specific responses in the antennae of the moths. A moth's antennae are attuned to the aromatics wafting toward it, and it may learn that a particular perfume, a blend of floral compounds, will lead toward a favorite reward. Hawk moths come into the world with a set of innate responses to, for example, linalool in wild tobacco, that are modulated through their complex antennae. The initial response of a naive moth is simple attraction to the molecule that is then modulated by experience and reward. Sort of an "Ooh, this is a good smell" followed by "This particular scent means a flower that has a reward." The ability of moths to learn different blends is dynamic, individual, and molded by a history with each flower species.[5] During nightly flights, moths learn to home in on the individual scent of a flower that is just ready to give up a bit of nectar or pollen in exchange for the receipt of pollen from another plant of the same species. The overall design of these flowers is such that the fuzzy body of the moth must brush up against the pollen-bearing anther as it gathers nectar from the long neck. A moth will use the same position to sip again and deposit pollen at the next flower.

Coyote tobacco (*Nicotiana attenuata*) grows in fire-driven desert ecosystems of Utah in the United States and is pollinated by sphinx or hawk moths in the genus *Manduca*. In-depth research into the relationship between the two has been conducted for over twenty years and provides two interesting stories of attraction and repulsion based on nicotine, fragrance, and halitosis. Aside from the flower, parts of the tobacco plant produce chemicals to serve a variety of purposes, and those found in the foliage tend to be protective in nature—repelling insects rather than attracting them. Nicotine, a type of alkaloid rather than volatile chemical, is a biological insecticide that is produced in the roots of the plant and distributed to foliage and nectar to protect from damage-causing caterpillars and other insects. Release of nicotine in wild tobacco is higher during the day, when chewing insects are more active. Some caterpillars, particularly in the family that contains hawk moths, are able to tolerate very high levels of nicotine; they can feed on tobacco plants and thereby gain protection from predators due to high levels of nicotine in their bodies. A few years ago, scientists were studying the relationship between *Manduca sexta* caterpillars feeding on coyote tobacco and nicotine in their diet to understand the relationship among nicotine, caterpillar, and predator—a wolf spider. Rather than keeping a toxic amount of nicotine in their gut, caterpillars transport it to their hemolymph, the insect equivalent of the bloodstream. When attacked, the caterpillars exhale small amounts of the nicotine through their spiracles (think nostrils, only in the abdomen) to repel the spider attack. And it seems to work. Through a tricky series of genetic manipulations scientists were able to modify the genes for passing nicotine between caterpillar gut and hemolymph, keeping the nicotine in the gut. More of the genetically manipulated caterpillars disappeared at night, when wolf spiders hunt their prey, leading scientists in the study to confirm the protective nature of nicotine exhalations against spiders. Authors of the study refer to this defensive mechanism as "nicotine-rich halitosis."[6]

Coyote tobacco has two faithful pollinators, the tobacco hornworm moth (*Manduca sexta*) and five-spotted hawk moth (aka tomato hornworm, *M. quinquemaculatus*), both of which are attracted to benzyl acetone. These moths, in addition to feeding on nectar and pollinating the flowers, lay their eggs on the plant to hatch into caterpillars that eat the leaves. Too many caterpillars are obviously not good for the plant, so when the plant perceives that caterpillars are munching on it (it can sense the oral secretions of the tiny, chewing caterpillar mouthparts), the plant secretes a compound called jasmonate. Jasmonate induces the plant to reduce production of the important moth attractor benzyl acetone and to begin flowering during the day. This tricky shift in fragrance and blooming time means that tobacco flowers are pollinated by hummingbirds, which don't eat plants, thus reducing the chance for moth pollination and caterpillar infestation.[7] At the same time as the chewing signal is going to the flowers, green leaf volatiles are sent out into the air from the damaged leaves to attract predators that feed on the eggs and young caterpillars. (Perfumers use these same chemicals to achieve a leafy and stemmy effect in perfumes.) Somehow the tobacco plant with its fragrant flowers has mastered the art (or is it science?) of attraction and repulsion.

Humans have a history of finding and using plant compounds to excess for their psychotropic effects, and nicotine is one of the most infamous. Perhaps cultivated as early as 5000–3000 BC in the Americas but also available growing wild, it was certainly smoked but also used by snuffing or inhaling powdered leaves and was even used as an enema. Where it grew locally, it was used to cure ailments as diverse as asthma (funny enough), rheumatism, convulsions, snakebites, intestinal disorders, coughs, skin conditions, and childbirth pains. Although tobacco was likely also used for pleasure, many records of early European explorers in the Americas recorded the smoking of tobacco as a ritual with religious or social implications—as with incense, the smoke of tobacco rising to the heavens would convey messages to the sky world where the gods resided. Tobacco has been a sacred plant to many Native American peoples, including the Navajos, who count it among their four sacred plants—the others being corn, beans, and squash—and use it in both blessing and healing ceremonies. Native tobacco, probably Aztec tobacco, or *Nicotiana rustica,* could be mixed with other herbs to make a blend with the Algonquian name kinnikinnick that was used ceremonially. Once it arrived in Europe, tobacco quickly became known as a miracle herb, was used for many ailments, and went by various names, including God's remedy and the holy herb. One of tobacco's first importers was the French ambassador Jean Nicot, source of the word *nicotine,* who experimented with it for a variety of ailments, even claiming that it cured cancer, and earning the plant the label "The Ambassador's Herb." Taking an example from Native Americans, Europeans began using a tobacco smoke enema to save drowning victims. The thinking was that the act would warm the drowned person and stimulate respiration. A simple pipe for blowing smoke up one's rectum by the rescuer was soon replaced by a device that contained a bellows, a pipe, and a variety of tubes to avoid the obvious dangers of backwash to the blower. The practice would go on to be used for a variety of ailments, including cholera, before being discontinued.[8]

From the first attempts at commercial production in the Jamestown Colony in Virginia to farms throughout the southeastern United States today, tobacco has been an important cash crop. In the early seventeenth century, colonists, especially in Virginia, devoted much of their time and energy to the production of tobacco, leading Charles I to say that it was a colony "wholly built upon smoke"—so much so that farmers had to be encouraged by the colony administrators to grow more corn for food. Earnings seemed large, but growing tobacco required substantial acreages of virgin land, someone with experience and skill to run the farm, no small amount of luck, and a large workforce. The need for labor in the fields was met in three ways: by having large families, by importing indentured servants from England and Ireland, and by purchasing enslaved Africans from slave traders. Tobacco plants grew readily in virgin land with a minimum of clearing; farmers would burn or girdle trees to plant tobacco among the dead trees and stumps.

Another tobacco derivative, neonicotinoids, have become some of the most widely used pesticides in the global market. Also called neonics, they are a collection of neurotoxins based on the nicotine molecule. At their introduction in the early 1990s neonics seemed safe because they could be used at lower levels than many other traditional pesticides and they were specific to insects while seeming to have lower effects on mammals. But because they are water soluble, they move freely through the plant, making their way from treated tissues to nectar, pollen, and even to guttation droplets—tiny drops of water exuded from the tips of leaves that are a source of water for pollinators. A study in France found imidacloprid, a neonic, in 40.5 percent of pollen and 21.8 percent of honey samples collected. Even if not sprayed in a particular year, a nicotine-based pesticide is likely to be taken up from the soil by a plant and may show up in nectar and pollen. In addition to spraying on crops, neonics are used as seed treatments in which seeds are coated with the pesticide before sowing and as soil additions. Although seed treatment may seem to be a very site-specific way of using the pesticide, the coating on the seeds is taken up by the plant to appear in guttation droplets but also produces a dust that may form a cloud around the area being seeded. Bees simply flying through this dust have been killed or, in the case of sublethal doses, may bring the pesticide back to the hive on their body. The dust cloud may spread to nearby nonagricultural areas, polluting them as well. Aside from direct effects, it took a while for researchers to show that sublethal effects may also be deadly. Because neonics are neurotoxins, they impair learning, memory, foraging success, hive hygiene, predator evasion, and other cognitive functions in insects. Honey bees have received a lot of attention, but bumblebees seem to be particularly susceptible. Despite the increased attention being paid to the dangers of neonicotinoids, they are slow to be taken off the market. However, in 2018 the European Union banned three of the most concerning neonicotinoids, clothianidin, imidacloprid, and thiamethoxam, for use in all outdoor applications.[9]

There are so many images of tobacco in the public consciousness— soldiers smoking during wartime, Humphrey Bogart rolling his own cigarette in *The Maltese Falcon*, Audrey Hepburn in *Breakfast at Tiffany's* with her fanciful long cigarette holder, and we all know what Freud said about a cigar. The Marlboro Man was a figure of romance and manliness, as was Clint Eastwood in the spaghetti western *The Good, The Bad, and the Ugly*. Smoking cigarettes has often been a symbol of rebelliousness and a way to flaunt authority in the public media à la

James Dean with cigarette and motorcycle. It is also a sometimes overt and sometimes more subtle symbol of sex—whether it is the act of holding hands while lighting a cigarette, sharing a puff, or simply slipping a cigarette between pursed lips as in the old movies. More recently, as I was involved in a complicated computer-driven project, I would sometimes hear a colleague (who had quit smoking) say, "I could figure this out if I could just have a cigarette!" Is it the ritual or the quick hit of nicotine?

Because the flower is not at this time extracted for fragrance, there is not a floral tobacco extract. However, the solvent-extracted absolute of the leaves provides a beautiful fruity-leathery base note to build upon. Perfumers have found the note to be unique and valuable, whether they use the actual tobacco extract (available nicotine-free) or similar compounds like clary sage or ambergris to obtain a tobacco note. An early tobacco-themed perfume, Habanita, was originally formulated to scent cigarettes and contained no tobacco, and Tabac Blond was introduced in 1919, just in time for the Roaring Twenties. The perfumes reflected a time when women were shedding corset, bustier, and bustle, bobbing their hair, smoking cigarettes, and wearing trousers. Rather than smelling like flowers (and their mothers), they chose a perfume that was bold and different. No longer advertised, banned in public places in many countries, carcinogenic, and addictive, tobacco products are, even so, abundantly in use today, telling us much about the fatal attraction and repulsion of this lovely and fragrant plant. As for me, I will take my tobacco in perfume form, thank you.

Herb gardens may be centered on fragrance as well as appearance and are also often laid out in lovely and interesting patterns. The stars of an herb garden include lavender, rosemary, basil, oregano, thyme, sage, tarragon, marjoram, and parsley. Although, like spices, they can be dried, stored, and shipped, herbs generally are best when used fresh and snipped from our gardens to be used in cooking but also for healing. Many common herbs are native to the Mediterranean biome, a fire-driven system characterized by moist and cool winters and hot, dry summers, weather that is often driven by wind and ocean currents. Rather than growing tall and leafy, plants in such habitats often stay close to the ground, keep their leaves small and tough, and protect themselves from herbivores by producing volatile compounds in leaf and stem. These terpene-rich herbs add depth and complexity to our foods.[10] The fire-driven ecosystem of the Mediterranean, or maquis, is one of several throughout the world, including the coastal California chaparral, southern Africa's fynbos, central Chile's matorral, and the Kwongan vegetation of southwestern Australia.

In between the tough aromatic shrubs and quick-sprouting grasses of Mediterranean systems you may find a glorious assortment of geophytes—plants with underground (hence the *geo-* in *geophyte*) bulbous, tuberous, rhizomatous, or corm root structures. Bulb plants are diverse and well adapted to these sometimes harsh and unpredictable systems driven by fire. Underground bulbs store nutrients and may have contractile roots that can pull the plant further underground as needed, perhaps to avoid drought, fire, or bulb-hungry herbivores. From the abundant orange poppies of the *Calochortus* genus in California to the beautiful amaryllis, hyacinths, orchids, and irises of southern Africa and daylilies, orchids, and even sundews of the Kwongan region in Australia, these plants share certain characteristics. To thrive they have developed the ability to store nutrients underground, flower and sprout when other plants are dormant, avoid hot, dry summers, regrow after fires, and draw in pollinators with their fragrant blooms. If, by means of their stored energy, they are the first ones to flower, they can attract pollinators with less competition from other flowering plants and use their fragrance as an additional draw. In southern Africa, pollinators may be long-tongued bombyliid bee flies (Bombyliidae) or tangle-veined flies called nemestrinids (Nemestrinidae) with extremely long tongues rather than bees or moths. One tangle-veined fly in particular, *Moegistorhynchus longirostris,* is a keystone pollinator in the Fynbos, where it pollinates at least twenty species of Iridaceae, Geraniaceae, and Orchidaceae that bloom in late spring or early summer. The flowers it pollinates have white or pink/salmon-colored flowers, are scentless, provide nectar, and have a very long floral tube well suited to *M. longirostris,* which has the longest proboscis relative to body length of any insect. Because these flies may pollinate several flower species, the flowers have evolved to deposit their pollen on different locations on the fly's body to avoid pollen contamination between species. These vibrant and unusual plants and their pollinators share ecosystem characteristics with some of the most common herbs and flowers, including lavender.[11]

Like a sommelier at a wine tasting, the lavender distiller carefully decants samples from sixteen types of essential oil from plants ranging from a lovely, pink-flowered variety to a somewhat strong and camphorous spike type. She has prepared them for my sniffing pleasure, and anticipating a peaceful session, I settle in to see if my nose can pick out the herbal, floral, woody, and sharp. Like many plants, lavender is a product of its terroir but also of breeding and the progression of the seasons. Although we may think of it as a floral fragrance, like many Mediterranean plants, it is mainly herbal in nature. I find the overall fragrance of most lavenders to be fresh and a bit sharp at first, but the floral linalool molecules present in some types shine through as the scent progresses on my skin or on a scent strip. There may also be a lingering note of floral woodiness that stays for a while both in a perfume and in an aromatherapy blend. As I sniff the carefully curated collection of lavenders, going first to various cultivars of English lavender, or *Lavandula angustifolia,* I find a varying combination of floral and clean, notes that are sometimes sharp and sometimes woody, with the occasional rooty or dirty note (but not in a bad way). Spike lavender, *L. latifolia,* is sharply green with a woody eucalyptus note, and *L. stoechas,* sometimes called French or Spanish lavender, is quite different from the others, with a somewhat incense-y note and a rough and woody aspect underlying the floral.

While most of us recognize flowers as the source of the essential oils, distillers know to gather top leaves and stems as well. Linalool is a monoterpene alcohol that provides the pretty floral scent from lavender and other flowers, where it serves to attract pollinators, but it may also be present as a protection in the leaves, where it can be released when the plant is chewed on by herbivores to attract predators that help control the little chewers. Linalool is a common and often dominant terpenoid component in many flowers, including lavender flowers and essential oil.[12] You may not have heard of it, but it is a sure bet you kind of like it: as do beetles, bees, moths, and butterflies. From everyday garden flowers to many of the fragrance products on the market, the presence of linalool adds a light and floral touch. Tea, fruits, roots, bark, citruses, wine, fungus, cannabis, and hops all contain linalool. Context is important with linalool: it is attractive in flowers but a weapon when the green parts of the plant are involved. It is emitted as part of a defensive arsenal of volatile chemicals that attract parasitic wasps and predators to prey on the chewers, a response that comes from the vegetative tissues of the plant, not the flowers. Linalool may be a component of sex pheromones in some plant-eating insects. And, interestingly, researchers collected urine from a wolverine that had hints of linalool along with other terpenes that were likely gathered from its diet of conifers and, unusual in mammals, passed through unchanged to the urine. Linalool can be isolated from plants like rosewood or basil by fractional distillation but carries with it traces of the parent plant, giving linalool ex (meaning from) basil an anisic hint while linalool ex bois de rose has a lovely and woody accent. Synthetic linalool is the primary form on the market; it is simply linalool without any scented baggage and may be produced as a product of the vitamin industry.

There are about thirty-two species of lavender that are generally aromatic shrubs with hairy leaves and stems.[13] While *Lavandula angustifolia*, English lavender, is the most desired species of lavender, producers also grow and distill hybrid and spike (*L. latifolia*) types. More than a thousand tons of various types of lavender essential oil are produced yearly. *L. angustifolia* is endemic to France, Italy, and Spain in the local limestone-rich mountain ranges up to about five thousand feet elevation. However, the plant is now grown in a variety of forms and cultivars around the world, including European countries, Australia, and the United States, and is widely used as a landscape plant for its attractive foliage and purple flowers. Lavender has been used for many centuries, being mentioned by Dioscorides in the first century AD. The generic name, *Lavandula*, comes from Latin *lavare*, meaning "to wash," and, in keeping with its name, was often planted near the laundry, where it was a familiar additive to the wash and often folded in between linens. It was also one of the herbs strewn on the floors of medieval homes to freshen the air and has long been associated with cleanliness and purity. Hildegard von Bingen wrote of two types of lavender, spike and what she called *vera* (probably *L. angustifolia*), for use in clearing the eyes and she recommended sniffing the aroma to terrify malign spirits. As with many of us, she likewise recognized its benefits in achieving a good night's sleep by bathing with lavender after a walk before bedtime. I recommend sprinkling a few drops on a pillowcase or handkerchief to help send yourself off to dreamland (with any luck, away from malign spirits). For ingesting lavender, Hildegard recommended blending with galan-gale, nutmeg, githerut (possibly black cumin), and lovage and then adding female fern and saxifrage and pulverizing well. The blend can be eaten on bread or compressed in a sort of pill. Lavender is often mentioned by medical practitioners throughout the centuries for its sedative as well as cleansing properties. Romans, a people who loved their baths, may have brought it to England, where it is an important landscape plant. Victorian England loved lavender, where it was an ingredient in early cologne formulations and was soon taken up by the House of Yardley in what was to become a long association. By the early twentieth century it was recognized for its antiseptic qualities and was used with sphagnum moss to treat wounds during World War I. Lavender essential oil has gone on to be the center of a resurgence of aromatherapy, or the use of plant oils for their emotional and physical benefits.

Pollinators, again, provide humans with the soothing and relaxing fragrance of lavender, but only at certain times in the life cycle of the plant. Plants emit volatile organic chemicals or VOCs from nearly all tissues but especially leaves, calyxes, and flower parts. The message the volatiles send is different depending on context and receiver of the message. Lavender plants have abundant scent glands called trichomes on their leaves, stems, and bracts that are likely protective in nature, whereas aromatics that the flower sends out may also be protective but are more likely for the purpose of attracting a pollinator. For the flower, the optimal time to send out its attractant VOCs would be both when the flower is open and receptive and when pollinators are active. Some flowers are quite able to do this, having evolved with their pollinators. A study of six lavender (*Lavandula latifolia*) cultivars grown in France found that there were three groupings of volatiles emitted at the bud, open flower, and wilted or pollinated flower stages that were protective in the bud stage, attractive to pollinators when emitted by open flowers, and then protective of the seeds after the flower wilted. Through control of activation and inactivation of enzymes, the lavender plant turned construction and modification of terpenes on and off in the flowering spike.[14] As with pollinators, people prefer the full-flowering stage where insect-attractive compounds are released, and in fact, international standards for lavender essential oil require a certain percentage of both linalool and linalyl acetate that the flower produces. Distillers have found that flowering stage, rainfall, and temperature all affect linalool in the lavender plant—environmental and plant parameters that also affect pollinator presence and availability.

Rosemary (*Salvia rosmarinus*) flourishes in the arid and chalky soils of the Mediterranean, growing from sea level to mountainside but also in gardens everywhere for decoration and culinary uses. Leaves and tops may find a place in a jar of vinegar alongside other Mediterranean herbs or may be added with some red wine to a marinade. We grow rosemary in our gardens and clip a bit here and there for the fragrance or to cook with. It may make an appearance at both weddings and funerals and was one of the herbs to freshen medieval homes. Churches, once upon a time, would burn rosemary for incense, and it was particularly associated with disease and loss of life during the years of the Black Death. Many of us know the saying "Rosemary for remembrance," but if you happened to live in the early fourteenth century in Europe, you might associate the smell with death and disease. The bubonic plague was stampeding through cities accompanied by the foul miasma of decay and uncleanliness—this at a time when pleasant fragrances were forces of good that promoted health while stink meant illness and evil. Wealthy people could leave the cities and surround themselves with fresh, clean air in their private gardens and well-ventilated homes, and they carried fragrant pomanders when they had to mingle, but the poor and ordinary folk were left with the medical tools they knew and could afford. Sweet- and fresh-smelling herbs had served them for centuries as floor coverings, food additives, and medicines—all good things. Rosemary was familiar, it was a part of life whether wedding or funeral, and it had a strong and cleansing fragrance. So, to fight the plague people carried twigs of rosemary, they added it along with spices and other herbs to vinegar or wine, and they burned sprigs as incense to cleanse their homes. If you can imagine the close and unsanitary quarters of a typical large city with open sewers, laundry on the lines, garbage in piles, and horse and other livestock droppings, and then add the stench of the dead in a city where people died faster than they could be carried away, you get an idea of the stink of the plague. Miasma is almost too kind a word. You can then imagine how the fresh and cleansing scent of rosemary, perhaps blended with some orange peel or inexpensive herbs such as lavender, sage, or thyme, could bring a bit of goodness and comfort to homes where the sick lay dying.

Rosemary, like other Mediterranean herbs, produces small flowers to attract pollinators, mainly bees, during the spring and summer. In the dry and varied topography of its habitat, patterns of bee activity change during the day, and different kinds of bees may be seeking different things—honey or pollen—that vary in presentation and availability. Because it grows at varying altitudes, rosemary has evolved the ability to alter the size of flowers, producing larger flowers at higher altitudes and smaller ones near sea level, while the tough, resinous leaves, referred to as sclerophyllous, remain the same small size.[15] This variation in flower size does two things: it results in flowers more accommodating to the larger-bodied pollinators of the mountains, and it protects them against the hotter and drier climate of lower elevations. Pollinators at higher elevations are likely to be the larger bumblebees that can thermoregulate to warm themselves up and therefore be active at lower temperatures than other bee pollinators. Leaves on the same plants do not have the same pattern of size variation; they are small and resinous throughout the range, making them resistant to desiccation. The standard list of constituents found in distilled rosemary includes 1,8-cineole, camphor, alpha-pinene and beta-pinene, borneol, and sometimes verbenone. Variations in the constituents, such as more cineole or more verbenone, affect both fragrance and mode of action and are labeled as chemotypes. Rosemary ct (for chemotype) verbenone is a good all-round rosemary type, and I like the woody and green notes that accent the typical resinous rosemary scent, whereas rosemary ct cineole, with its herbal sharpness is a slightly different take on rosemary.

Rose (Rosa rugosa) *with hip*

8

Roses

As she tasted the rose lassi, my granddaughter had a look on her face that was a little hard to interpret. I asked her if she liked it, and she was not quite sure. It was her first time eating at an Indian restaurant, and she enjoyed the entire meal but continued to hesitate about the lassi. I admit to helping myself to a few sips, and it was just as I had hoped, full-bodied rose over rich yogurt, slightly sweet, and a treat for both nose and mouth. But then I love a full-bodied rose extract and have learned to make my way through the complexity of scent. My granddaughter spent a little time that evening sort of coming to terms with the conjoined impact of scent and taste, floral and tart, and I hope she will remember the experience fondly. A rose lassi, as with other Middle Eastern dishes, is made with rosewater and rose preserves.

Beginning with Linnaeus there has been much frustration about how to classify roses and how to determine the origins of our domesticated varieties. Roses have a habit of hybridizing freely that results in a lack of clarity for scientists and in enthusiasm for rose breeders. There are about two hundred species of rose, most of them wild species with elegant five-petaled blooms in white, pink, and red, backed by five sepals and a variable number of bright yellow or orangish stamens in a distinctive circle in the middle of their blooms, the better to attract pollinators. Prickles or thorns vary in size and number. Most rose species seem very adaptable to sharing and incorporating genes between species and between cultivars, habits which have resulted in the variety of roses we see in gardens and flower shops today. Two paths have existed in rose breeding that led to the cut-flower rose bred for color, shape, and longevity in the vase and to roses grown for gardeners who appreciate scent as well as beauty. Although it is instinctive to bury one's nose in a bouquet of deeply colored roses, it is almost always frustrating if you have purchased them as cut flowers, whether from a grocery store or a high-end florist, and not gathered them from your garden. Commercially cultivated flowers have little or no scent, and scientists actually do not know exactly why since the genes and enzymes are present: there is simply no follow-through by the rose to emit scented molecules from the flower.[1]

Through much of Western history, as we saw with the plague, scent was a force for goodness—and then it was not. In her book *Worlds of Sense,* Constance Classen notes that smell is a neglected sense in the West, being taken over by sight. Somewhere around the eighteenth century, attitudes in Western cultures began to emphasize the visual: odors were minimized by sanitary measures in cities and by various deodorizing products. These attitudes were reflected in both gardens and the breeding of flowers. Gardens became a visual experience, and flowers were bred for color and longevity, neglecting their gorgeous fragrance and giving us the modern scentless cultivated rose.[2]

We will leave those scentless roses in their bins at the market and head back into the garden for both beauty and scent. With an abundance of rose species and types, some division into general groups may help in a discussion of rose types and species. There are types that flower just once a year and include albas, centifolias, damasks, gallicas, and moss roses, while repeat flowering types include Bourbons, Chinas, hybrid perpetuals, noisettes, Portlands, and tea roses. There are climbers and shrubs, thorny and not, wild and cultivated, single and many-petaled. A few iconic and important ones deserve mention.

Rosa banksiae flowers early with a violet-tinged scent: it is a single bloomer named after Sir Joseph Banks and originating in China. A cutting was brought to Tombstone, Arizona, in 1886 by a Scottish resident and planted in the yard. Soon it required support in the form of a trellis, and it is still growing today with a circumference of about eleven and a half feet and a five-thousand-square-foot canopy, give or take. The house became a hotel called the Rose Tree Inn and is now a private residence, but the patio is open to the public. *Rosa × centifolia* is the hundred-petaled rose and blooms once with a sweet and strong fragrance. The beautiful blooms have been featured in Dutch flower paintings from the seventeenth century, and it is valued for its large flowers and sweet scent. Moss roses belong to this group and are a mutated form with aromatic glands that develop on the sepals, calyces, and pedicels taking the form of a mossy growth that is highly aromatic and sticky, smelling of pine.[3]

Rosa chinensis, Chinese rose, has been cultivated in China for probably two thousand years and includes the development, about a thousand years ago, of the cultivar Old Blush, which gave its genes to modern roses, making it possibly the most important rose in the development of modern types. Old Blush contributed the continuous flowering trait to modern roses and was brought to the West in the eighteenth century, where it was crossed with European roses like damasks and gallicas to get Bourbon and hybrid perpetual roses. The scent is medium strength and tealike; it is a continuous bloomer with pink double flowers. The only green rose in existence originated as a mutation and is called Viridiflora. *Rosa × damascena,* or the damask rose, is a hybrid of *R. gallica* (apothecary rose) and *R. moschata* (musk rose) and one of four important sources of rose oil for the fragrance and flavor industry. *Rosa foetida,* the Persian yellow rose, has a strong and fetid scent, comes from Asia, and has bright yellow flowers. A sport called Persiana was used to breed the first truly yellow garden rose, Soleil d'Or, in the year 1900. The stinky scent attracts flies, wasps, and beetles.

Rosa gallica has a sweet "old rose" scent and blooms once. One of the most important of the wild species, it is an ancestor to almost all modern roses and comes from France and central Europe, Ukraine, Turkey, and Iraq. Sepals and receptacles have a sticky scented resin. The Red Rose of Lancaster is an early cultivated variety of *R. gallica* that may have arisen in France from the wild species in about 1400 and was grown for medicine, perfume, and preserves as well as for its beauty. The Rosa Mundi sport, also called Versicolor, has white stripes and arose in about 1560 in Britain. *Rosa moschata* is the musk rose, an ancestor of the modern rose, a single-flowered species with a hint of musk to its strong rosy scent. It may have originated in the western Himalayas and has since naturalized throughout the Mediterranean region. *Rosa multiflora* has a strong and musky scent and is one of the most important wild rose species for use as rootstock. But more important, it is an ancestor of the multifloral ramblers, the polyantha roses, the floribundas, and the grandifloras. The flowers are small and white, but there may be up to five hundred on an erect panicle. It is native to northern Japan and parts of Korea.[4]

Perfumers use just a few of the many types of rose: damask rose (*Rosa* × *damascena*), rose de mai (*R.* × *centifolia*), the white rose of York (*R.* × *alba*), and a hybrid of the Turkish rose (*R. rugosa*) are the main roses used in the fragrance industry today, and they are grown in Bulgaria, Turkey, Morocco, Iran, Afghanistan, China, and India.[5] The aroma is extracted by distillation to yield the beautiful and precious rose otto essential oil together with the aqueous portion called rosewater. Alternatively, solvent extraction produces a solid, waxy mass called a concrete that is further extracted with alcohol to produce absolute of rose, the form most often used in perfumery. Absolute of rose de mai, rich in phenyl ethanol, has a deeply rosy fragrance that is less spicy than damascena. Damascena absolute is rich and somewhat more muscular, being warm and spicy with very rich rose tones.

Distillation of the rose, almost always damask roses but occasionally rose de mai, generally involves the entire flower, carefully plucked in the early morning hours before the sun rises too high. What I would like to do here is go back about 150 years to an old document produced by growers in the Kazanlak area of Bulgaria.[6] They describe fields of red damask roses growing in dense hedges separated by rows wide enough for two men to walk abreast. The red roses were surrounded by white-flowered rose bushes, forming a border, and the pickers would enter the field in the early morning and pick only open or half-open flowers with the dew still on them. At the time they estimated that each acre produced more than a million flowers, requiring about eight hundred person-hours to produce 6,600 to 8,800 pounds of petals that yielded just over 2 pounds of the distilled rose otto, an estimate that is comparable to production today, although highly managed fields and improved cultivars have increased yield per acre. In addition to the race against time, in the traditional fields there was also the calculation to match the rose harvest with available distillation units. A small distiller could be overwhelmed with a good harvest and, if unable to distill the flowers quickly, would find it necessary to sell to larger distillers. Wood to heat the still and water to suspend the flowers as well as to cool the condensation unit were also required. Production of rose oil is still somewhat based on small family plots that send their roses to a larger distillation facility, meaning that the industry supports a number of small family farms.

When an aromatic plant such as the rose is distilled for its essential oil, two things are produced. One is the oil, which floats on top of the liquid in the receiver and is called the otto. The other is the aqueous portion (rose water in this case), often called the hydrosol, hydrolat, or floral water. Hydrosols contain the water-soluble alcohols and other aromatics from the plant and often tend to be more lightly scented and gentle. Rose absolute is the other aromatic product used by the perfume industry and is produced by solvent extract of the flowers. Extraction by solvent first produces the concrete, which contains floral waxes from the blooms along with the many lovely aromatics. The concrete is further extracted with alcohol to yield a dark orange-red absolute that is richly rosy and soluble in alcohol, making it easier for perfumers to use. Although little used, I have found the concrete of roses to be exceedingly lovely. The solid waxy lump may be a bit difficult to work with, but it holds the scent for years after purchase and can evoke the fragrance of the rose in a deeper and more complete sense than either the absolute or the essential oil.

Two volatiles contribute to the iconic fragrance of the rose. The scent of Chinese roses includes a molecule called DMT, or 3,5-dimethoxytoluene, that is the reason for their other name, tea rose, which refers to the fruity tea notes of their flowers. European roses, on the other hand, produce and emit 2 phenethyl alcohol (2PE or PEA), which is honeyed and richly, deeply rosy. The overall "rosy" scent of rose is modified by various other aromatics to give each type a unique aroma. In addition, the different parts of the rose flower produce different types of volatiles depending on the purpose. To make the scent of a rose requires as many as three hundred constituents to provide the complex and yet recognizably rose fragrance of the flowers. Most roses also give us the floral, the spicy, the green and fresh, citrus, and even the scent of myrrh and precious resins, fragrances that come from other constituents of the highly complex volatiles produced by the rose flower. In addition to DMT and 2PE, rose ketones—beta damascenone and beta damascone—give the flowers a fruity floral odor, and rose oxides provide the green and spicy. Rose ketones are important to the fragrance industry to build a rose fragrance from constituents, and rose oxides occur in two forms depending on optical rotation of the molecule—the cis has a sweet and floral fragrance with a distinct green note, while the trans adds a spicy aspect. Some roses have a hint of methyl eugenol to provide a bit of spice.[7]

The simple flowers of a wild rose range from white through pale pink shading into dark pinks and reds, but all have a crown of bright yellow or orange stamens in the center. Petals produce a sweet fragrance, sepals are rich in sesquiterpenes for a piney, lemony scent, and anthers and pollen have a diversity of unique compounds that are not shared with the rest of the flower. These fragrances serve to attract specific pollinators to different parts of the plant. Wild roses like *Rosa rugosa* do not always produce nectar, but they have a ring of bright yellow stamens that produce an abundance of pollen to attract pollen-seeking insects such as bumblebees. A study of wild roses in Europe found variable numbers of stamens between 83 and 260, which is a lot of plant material for the rose to invest in and supports the idea that attracting pollen-seeking insects is important for at least some of these rose species. Bumblebees use both the visual cue of a bright ring of pollen-bearing stamens and the scent of the anthers, with their hint of spicy eugenol among other compounds, to home in on the protein-rich pollen.[8] *Rosa rugosa* is considered invasive in Europe and has a medium strong scent that is lighter than damask varieties and stronger than the dog rose (*R. canina*), a climbing species native to Eurasia and northwest Africa. Eugenol in the pollen seems to be a strong attractant for bumblebees, resulting in both increased landing and vibratile pollen-collecting behavior. Bumblebees are experts at vibrating, also called sonicating, to release pollen from various kinds of flowers; in doing so they produce that audible buzz so familiar to gardeners on summer mornings. Bumblebees mainly use this tool when collecting pollen from plants like blueberries, cranberries, and tomatoes that have pollen enclosed within the anthers to be released through small holes or slits: vibrating at a particular frequency releases the pollen. For roses and other flowers with abundant stamens, they seem to almost wade through the pollen-rich center, gathering the good yellow stuff, buzzing, and pollinating as they go in a behavior called scrabbling.

Although Europeans had a variety of beautiful and fragrant wild and cultivated roses, the importation of the Chinese rose in about 1780 brought the ability to bloom more than once, and breeders quickly added them into the gene pool. Chinese roses arrived at a time when there was immense interest in flowers. Breeding combined European disease- and cold-resistant varieties with repeat blooming, tea-scented Chinese types, giving life to a passion and industry built around the rose. Persians, who are responsible for the discovery of distillation, have a great love for roses and would include them in nearly all gardens, often in hedge plantings, to lend their scent to jasmine, lilacs, and violets in the beds. The city of Shiraz was home to legendary rose gardens of great profusion and fragrance. Persian warriors carried the image of roses on their shield, and the province of Farsistan paid an annual tribute of thirty thousand bottles of rose water to the treasury in Baghdad. *Rosa × richardii,* the holy rose, is rumored to be present, and still fragrant, in Egyptian tombs. The Wars of the Roses in England's fifteenth century symbolized the dueling sides with a red rose for the House of Lancaster and a white rose for the House of York. Cleopatra covered the floor of her royal apartment with roses ankle deep when she first met Mark Antony, and Romans wore crowns of roses to cure hangovers. Theophrastus described both cultivated and wild roses as well as the manufacture of perfumes by extraction into sesame oil. Dioscorides describes salting the petals to preserve them, Pliny had a formula for rose wine, and the Victorians ate rose petal sandwiches. Shiraz was distilling an essential oil by the end of the seventeenth century, and roses reached Kazanlak, Bulgaria, about the same time.

Before I leave the subject of gardens, a word or two about orchids. As a child of the American West, I loved the spare landscapes where one could see the geology and where flowers were seasonal, often requiring luck, spring rains, and careful searching to find. As an adult I moved to Florida and discovered an entirely different world where plants covered nearly any exposed space and orchids are happy to grow and thrive. I mostly learned to care for these flowers through trial and error and learned that they are quite tough but that there just may be some you cannot succeed with. A rule of thumb is that if you kill a particular type three times you need to give up and focus on those you can grow. The subject of orchids is as wide as the world since they are one of the largest groups of flowering plants and their distribution extends over all continents, except possibly Antarctica, and they are embedded deep in time, perhaps having coexisted with the dinosaurs. There are about thirty thousand species of orchids, and most of them accomplish fertilization by way of pollinators but through scent and trickery. These are their tools, and they use them well.[9]

Here I would like to tell three stories from my experience with orchids that relate to the wider world of orchid species. Orchids often attract pollinators through deception, a successful strategy that involves producing a flower that resembles the insect they want to attract and encouraging copulatory behavior that results in pollination. There is no reward for the insect, but the orchid may manage to deposit a little packet of pollen called a pollinia for the insect to deliver elsewhere. Some orchids, such as the *Gongora odoratissima* in my backyard, do offer a reward, and the reward is fragrance. Rather than spending money on the blooming one at an orchid show, I bought a young plant with two or three bulbs but no blooms. A couple of years later I closely followed the lengthening of a single flower spike hanging from the plant and waited for my little dragon-shaped flowers to appear. Sure enough, the dragons began to emerge, and I proceeded to visit the flower as each day brought new blooms. Then I noticed the light scent of cinnamon and enjoyed my visits even more. But the best surprise of all, well, almost the best, was the bright and shiny blue bee that showed up one day. A bit of research gave me its name. It was an orchid bee, and a bit of observation showed me the large hind legs characteristic of the group. I ended up placing a little stool so I could sit near the orchid to watch the bee and found that he (the scent gatherers are always male) would fly up to check me out, hover in front of me for a moment, and then go back to assume his position within the flower. In an act that looked very much like copulation, the bee was actually scraping scent from the petals of the orchid. After a minute or two he would fly a few inches away from the flower, and I could see him moving something from his forelegs toward the large pouches on its hind legs. Then back to the flower he would go. He was storing fragrant waxes from the flower in those pouches to be later blended into perfume and wafted from a prominent perch to attract females. I never saw any pollinia on the bee, but he visited the flowers as long as they were in bloom.[10]

Books have been written and movies made about the passion of orchid collectors. In my little part of the world, I have seen a small part of this passion. Each year, usually between January and May, the orchid shows arrive in south Florida, as I am sure they do in other parts of the world. Vendors from all over the world bring plants from tiny bulbophyllums to large vandas with their fleshy aerial roots, and vanilla vines tangled together in a cardboard box. It is best to arrive early and join the eager line of collectors, most of whom have brought little wheeled wagons to fill with orchidaceous treasures. I admire the courage of those who will spend big bucks on a beautiful flowering cattleya or rare blooming vanda. But with nothing particular in mind, I usually enjoy wandering and waiting for something to catch my eye. Below the showy blooms on multiple displays are the bare-root plants that are available for a fraction of the price but require patience since they often take one or two years to bloom. I am okay with that because I have found that anticipation is part of the fun and I do not feel quite so bad if they expire. Orchid collectors in the past were among the plant explorers of the eighteenth and nineteenth centuries who collected at will, often destroying entire populations of beautiful specimens to ensure their monopoly.

Tiny *Angraecum distichum* is one of my favorite orchids, and mine is about ten years old. With thick alternating leaves less than half an inch long, and a total length of under four inches, this little orchid fits nicely among the showier ones. It puts out new growth sporadically through the year, with tiny white flowers appearing from between the leaves. The flowers are fragrant but too tiny for me to get more than a hint of their sweetness. Another orchid in the same genus, *Angraecum sesquipedale,* has flowers almost as large as my hand. Called Darwin's orchid, the flower is famous for confirming his theories on pollination and pollinators, though not until after he died. When the orchid was discovered, the large white flowers were found to have a nectar tube about twelve inches in length, and Darwin theorized that there would be, somewhere, a moth with a tongue that would reach to the bottom of the tube to gather the nectar. A moth fitting just that prediction was eventually discovered.[11] This single genus contains fragrant, white-flowered orchids that range in size from just a few inches tall to over six feet in height. As I enjoy my orchids, I realize that I have a similar size range in the entire collection from tiny bulbophyllums to a couple of dendrobiums growing to three feet tall or more with abundant blooms. Orchid flowers may resemble a bird, a small human, a spider, or even a parrot, and some have bulbous bases that resemble testicles, giving the group their name, which is derived from the Greek *orchis* for testicle.

Our ties with flowering plants go back to our origins, and they are with us in life and in death, in homes and in gardens large and small. They have provided medicine and food, beauty and fragrance, and those flitting and buzzing pollinators, including butterflies, moths, bees, and hummingbirds, that value the weight and beauty of a petal. Their mystery? Darwin thought it was their origin, but I wonder if perhaps it is how did tobacco and sphinx moth become so close, where does the mushroom scent in gardenias come from, why is there not a blue rose, and did dinosaurs eat orchids?

PERFUMERY FROM MANDARIN TO MUSK

A book about fragrance would not be complete without the story of perfumery and the extraction of scented compounds for the purpose of beauty and attraction. Plants have mastered the art of creating and blending fragrant molecules using hundreds of individual scented compounds. Scent from a flower reaches out to moth, butterfly, bee, or beetle, enticing a passerby to come near, sip sweet nectar, and pick up a little packet of pollen for that other plant over there. Other parts of a plant also blend and send out scented molecules, but to attract predators of their enemies or to warn other plants of attack. Some aromatics are retained in a plant's tissue to heal and defend against disease. Humans have also used plants for healing and attraction, turning to aromatic plants to make medicine and perfume. The complexity of extracting and using aromatics from plants meant that, for most of history, perfumes were reserved for the very wealthy, often royalty, who kept their own alchemists to distill and create fragrance from flowers, woods, spices, herbs, and musk. Science and industry eventually became involved to prepare specific aromatics for perfumes. In the south of France, a city called Grasse grew in the limestone mountains near the blue Mediterranean Sea where lavender was found, jasmine thrived, and the industry of perfume had its beginnings. The story of perfume begins with medicine and alchemy and three fragrant accords.

Before that, early perfumes were taken directly from nature in the form of flowers, leaves, wood, resin, seeds, and roots and were extracted into fats and oils or dissolved in wine and vinegar. For most of our time on this planet, plants for medicine, food, and scent have been the same thing. Europeans drank cologne, agarwood has been used for mental and physical health as well as to scent clothing, ambergris was paired with rosemary to avoid falling to the plague, rose hips and jasmine buds make excellent tea, and both rosewater and orange-flower water have been extensively used in cooking. In or around 1370, a hermit or alchemist (the story is a bit fuzzy) gave Queen Elizabeth of Hungary a formulation called Hungary water. With notes of rosemary and citrus, it could be considered the first cologne and was meant to be taken in several ways: internally as medicine, rubbed on the skin for beauty, or used as a cologne. Napoleon appreciated cologne and used gallons of citrus cologne each month, in counterpoint to his beloved Josephine, who favored strong and lasting animalic musks.

Beginning with Greek scientists and passing on to the Arab world, alchemists studied distillation and strove to make the invisible visible. But that alone was not enough—the business of modern perfumery also required the ability to make both alcohol as a carrier for fragrant materials and fine glass vessels to hold the perfumes. With new scented ingredients arriving in Europe, glassmakers refining their art, and royal courts having their own alchemists, the transition from alchemy to perfumery began. Apothecaries, spicerers, and chymists helped to develop the industry of perfumery and brought scented products to the people.

In my small way I have replicated the process used in still rooms for centuries to make medicine and food but also in industry to produce essential oils for our perfumery, flavoring, and aromatherapy needs. Magic, or alchemy, was putting the fragrant leaves of a lime tree into the glass retort of my little still, covering them with water, heating the leafy mixture, and watching tiny droplets of steam rise to be captured in a long, narrow condenser tube. The droplets flowed down the sides of the condenser and filled a small jar below with two layers of fragrance. Heat and steam are required to separate fragrance from plant, breaking vesicles and plant matter down to release aromatic constituents that rise with the steam. Those tiny packets of steam contain both water and essential oil that turn to liquid droplets when they meet the cool outlet chamber of the condenser. These droplets run down the sides of the condenser and collect in a receptacle where they appear as two layers: essential oil and aqueous portion. The fragrant liquid floating atop the water was thought to represent the essence of a plant and so came to be called essential oil, from the old Latin *quinta essentia,* referring to the fifth element, or pure essence, of a thing. Below the essential oil is the aqueous part, called a hydrosol or floral water. When plant material is placed in water it is called hydrodistillation, whereas steam distillation involves pumping steam through the botanical material.[1]

An ancient process of scent extraction practiced by Egyptians was a basis for the art of enfleurage adapted by the perfume world for precious flowers that cannot withstand the heat of distillation and that continue to release their scent after being picked. Jasmine, tuberose, and hyacinth are some of the plants that do this, which means they breathe their pure scent into the air or, for the process of enfleurage, into a semisolid fat where the molecules are captured. If you have ever kept an onion next to butter in the refrigerator, you know that fats absorb odors. In Grasse, France, they took the process to the next level and used descented and purified lard spread on glass plates. Jasmine flowers, for example, would be picked first thing in the morning and quickly brought to the workshop, where agile hands laid them face down on the larded glass. After a day of exhaling into the fat, spent flowers would be removed and new ones put in their place: as many as thirty-six charges were used to fill the fat with scent. The result was a pomade, basically a scented fat, which might be called "pomade de jasmine no. 36" based on the material and number of charges. The fragrance would then be extracted from the fat using an alcohol wash to produce the "extrait de pomade no. 36." Since the alcohol-based extract was produced simply from the floral scent, with no heat or plant material involved, it was, and still is, considered the most natural reproduction of the flower's perfume. Once extraction with a solvent like hexane was developed, it replaced enfleurage as a way to capture the exquisite scent of jasmine or tuberose. Solvent extraction yields the solid called concrete that contains wax from flowers along with scented molecules and must be washed or extracted with alcohol to yield an absolute perfume oil, the form used for perfumery. Absolutes are powerfully fragranced and soluble in alcohol—a requirement for the perfume industry.[2]

Eastern white pine (Pinus strobus) *needles and cones*

9

Humble Beginnings:
Mint and Turpentine

Compared with scented flowers, mint and turpentine seem humble and unassuming, but both have been integral foundations of the fragrance industry and of the early economy of North America. Distillers of mint created first demand and then an industry as mint has become one of the top three selling essential oils. Turpentine is a starter material for many aromatic chemicals in use today and may come from a variety of conifers, including pines with abundant resin from the southeastern United States.

While distillers and perfumers in Grasse in southern France were focused on the highly favored florals that grew so well in their climate, distillers in North America found humbler crops on which to build their industry. Mint has been popular since earliest times and is a familiar dooryard herb, especially peppermint (*Mentha* × *piperita*), which originated in Europe as the offspring of two related mints, water mint (*M. aquatica*) and spearmint (*M. spicata*). Perhaps it was first found by gatherers in wet areas where water mint and spearmint grow together, producing a hybrid mint with traits surpassing those of the parents—at least in usefulness to their human gatherer. Because mints grow so easily from roots and cuttings, it was readily propagated and passed around. Early medical practitioners who used mint would chop the leaves into tiny bits to disguise its identity, but as with so many secrets, this one was found out, and soon bits of roots and stems were being smuggled, eventually making their way as cuttings from Europe to North America. In the Americas mint may have been used to make summer teas and mint juleps: mint vinegar was made with cider vinegar, mint leaves, and granulated sugar boiled and strained to flavor fruit punches and sauces. Candied mint leaves could be made by dipping the leaves in whipped egg whites and then coating with sugar. Fresh and cool, the flavor and fragrance were treasured in hot summer weather.

A popular item in the home herbal medicine chest, mint became one of the first distilled plants in North America and the foundation of an industry. It began in a small way near the centers of rum distillation and was soon being grown specifically for essential oil production in Connecticut and Massachusetts. Essence peddlers located near peppermint farms spread the gospel of peppermint essential oil. Ambitious young men who wanted to see the world and make some money set out during the mid-1800s with trunks or baskets yoked to their backs containing sturdy little bottles of mint, bergamot, bitters, and other bits and bobs for sale. As they made their way across New England and into Upstate New York, they earned a living and eventually assisted the spread of the essential oil industry into new territories. Among other visionary men, the chemist Albert M. Todd began producing peppermint essential oil in Kalamazoo, Michigan, in 1891, and his "crystal white" mint made him one of the peppermint kings— and the company he founded is still in business selling mint products. Mint is one of the most widely produced and used essential oils for food and products such as toothpaste but also as a source of menthol. Oregon and Washington are now centers of mint production in the United States. Corn or field mint (*Mentha arvensis*) grows wild in China and Japan and is cultivated for the abundant menthol in its leaves. In England, Mitcham peppermint oil or black Mitcham oil is grown and distilled to make a rich and sweet essential oil, complex and full of body, that is used mainly in high-end products.[1]

Clary sage (*Salvia sclarea*), a native of Mediterranean habitats, is the source of a popular essential oil, but the plant is also grown in conjunction with tobacco as a flavoring. Sometime in the 1950s an employee of R. J. Reynolds, a North Carolina tobacco company, found that the extract of clary sage added a desirable aspect to cigarettes that could help re-create the flavor of milder imported tobacco, giving the company the ability to control price and quality. In 1958 R. J. Reynolds got a patent for using clary sage in cigarettes to enhance the aroma and flavor of domestic tobacco. The company also began trial production of clary sage in eastern North Carolina, where it is still grown today both for the tobacco industry and for a chemical by-product called sclareol, which is a precursor in the synthesis of some of the hugely popular musk aromatics used by the perfume industry.[2]

My first experience with pines of the southeastern United States was in South Carolina. We were headed out in a couple of four-wheel-drive vehicles to band swallow-tailed kite (*Elanoides forficatus*) youngsters on the nest in a longleaf pine (*Pinus palustris*) forest as part of a research project. As soon as we got out of the car, my coworker handed me a roll of duct tape and showed me how to tuck my pants into my boots and tape them off: chiggers and ticks abound in the tall grass of the longleaf pine ecosystem, and the duct tape at least acts as a deterrent, if not preventative. As a newcomer to the pinewoods of the South, I was unaware of this creepy danger. At the very tippy tops of the tall, slim pines were the kite nests, and I watched as adult birds circled and called while another coworker climbed a tall tree with the aid of tree spikes and carefully roped-in harness to bring the youngsters down for banding. For those who have not seen them, swallow-tailed kites are one of the most beautiful and graceful of the raptors and are hunters of flying insects and small animals of the piney woods. This once-in-a-lifetime experience was well worth the risk of insidious crawling insects, and I soon learned the art of duct-taping my pants, keeping on the move through tall grass, and watching for fire ant nests (a whole 'nother danger of the South) as I worked on projects in the pinewoods. But with these irritations came the opportunity for rare beauty as I listened to the many bird calls, watched woodpeckers and kestrels make their homes in the tall trees, and found the occasional pair of wood ducks nesting on small ponds tucked into the woods. You can tell where red-cockaded woodpeckers (*Leuconotopicus borealis*) nest because they know to find older pine trees with the most heartwood for their nesting cavities. Excavation of the heartwood produces an obvious flow of resin to coat the tree under the nest and keep out such predators as rat snakes.

The resin that sticks to my hands reminds me of my walk in the woods for the rest of the day. Resin like this, along with wood from the trees, was an economic driver for the early American colonies and one reason for widespread loss of these very same forests. Turpentine was one of the first essential oils distilled in the American colonies from seemingly endless pine forests of the East Coast, and it continues to be an important source of fragrance and flavor materials. Beginning with the eastern white pine (*Pinus strobus*), with its tall, straight trunks, trees were harvested along the East Coast. In the colonial era, the British began large-scale harvesting of the abundant virgin forests for masts and other naval shipbuilding needs. As tall, old-growth individuals were removed, local authorities attempted to claim ownership of the largest for their naval vessels. One result was New Hampshire's Pine Tree Riot of 1772, a little-known precursor to the Tea Party Rebellion and American independence. In addition to providing tall, straight trunks for ship building, early colonists ventured into the deep woods to harvest a sticky resin from pine trees, including eastern white pine, slash pine (*P. elliottii*), and longleaf pine, to supply the naval stores industry (the Royal Navy's use of pine products) and the many wooden ships sailing the oceans of the world.

Settlers also harvested resin for medicines and flea repellents, as well as tar and pitch for their own use. For more than four hundred years turpentining took place in the immense pine forests of coastal North America. Gathering of resin or gum was conducted deep in the woods, where workers would live in camps that moved with the availability of trees, and workers were kept poor, in debt, and at the mercy of overseers. Resin was harvested by cutting through the bark of the trees and into the wood to collect it as it dripped down the face of the trunk into a wooden box. There was often no finesse and no consideration of the health of the forest—trunks would be deeply cut and damaged, leaving openings for insects and disease. Once an area was spent, turpentine camps were moved, and damaged trees left behind: in the early days they were not harvested for wood since there were abundant virgin forests all around. In addition to resin, turpentine and rosin may be distilled from the stumps of felled trees. Sometimes the pinewood was piled into a kiln and covered with earth to force out tar by means of a slow heat. Working these stills meant walking in the tar and gave rise to the nickname "tarheel" for North Carolinians, residents of the Tarheel State. Today, anyone who has traveled or lived near a paper mill may recognize the funk of the sulfur kraft process of paper production, which yields sulfate turpentine. This form of turpentine is recovered from wood chips and is an important source of such synthetic fragrance ingredients as floral ionones, flavor ingredients, and vitamins A, E, and K by way of chemical changes to the constituent terpenes.[3]

Resins may squirt from the leaves of a copal bush to defend against little herbivorous flea beetles, ooze into exquisite drops on the trunk of a frankincense tree to heal a wound, or, in the case of pine trees, may drip and flow in response to bark beetle attacks. There is a highly orchestrated set of steps in the battle between bark beetle and pine tree in coniferous forests of western and southeastern North America. Bark beetles of a variety of species will settle on a tree and begin the attack, boring into the trunk, to which the pine tree responds by releasing liquid resin to repel the beetle. To overcome this defense, bark beetles emit an aggregation pheromone to call in reinforcements that also begin attacking. The tree releases more resin, and more beetles attack, until finally one side wins. Sometimes the tree repels the beetles, but sometimes the beetles win and begin burrowing in. Too many beetles will overwhelm the tree resource, so once they take up residence, winning beetles send out antiaggregation hormones that keep other beetles away. These homeless beetles may find nearby trees to begin the process all over again. Female bark beetles burrow into the tree, creating galleries under the bark where they lay their eggs for the larvae to hatch and grow into adults. Damage from the beetle and disease from accompanying fungal pathogens are often fatal to the tree. Both resins from the tree and pheromones from the beetles are heavy with terpenes, including verbenone, which has a camphoraceous, green, and celerylike fragrance.[4] Bark beetles also attack other conifers throughout the world, and potential for serious damage to the world's coniferous forests is increasing as the world warms, stressing the trees, and reducing killing-cold temperatures that help control the beetles.

Once covering an estimated ninety million acres throughout the Southeast, virgin longleaf pine may now cover only a thousand acres or so and second-growth forest another two million. Turpentining was not responsible for all this loss: conversion to agriculture, urban development, fire suppression, bark beetles, climate change, and timbering have also posed challenges to the piney woodlands. Pinewoods are fire-driven systems and require occasional fires to clear underbrush, control competing vegetation, and stimulate pine seeds to sprout. Longleaf pines spend years making their way from seedlings in the grass stage where they look nothing like a pine tree and more like a clump of grass while growing a large taproot to sustain the next stages. Next comes a sapling that is basically a long, thin trunk with a bottlebrush top of long pine needles, a structure designed to keep that top above any fires that may pass by. Then come the tall, majestic trees with free-running resin, red-cockaded woodpecker nests, and a diversity of wildlife in the tall grasses and ponds. Although you may see over five million acres of pine trees in South Carolina, for example, they are mostly working forests of loblolly pine (*Pinus taeda*) or slash pine (*P. elliottii*) planted in straight rows like a cornfield and harvested for wood products.[5]

Although most of us do not think of conifers as perfume ingredients, quite a few may evoke woodlands and forests, seasons and holidays, incense and resins, or just add fresh green interest to a scent. There is also the perfume of the outdoors where these trees grow. The Japanese practice of forest bathing, or *shinrin-yoku,* means simply spending time in nature, mindfully, without distractions or worries. You do not really have to bathe in a forest of conifers—neither immersing yourself in water nor walking under conifers is necessary—but conifers often provide the kind of stately peace that seems appropriate for such an activity. It seems that pretty much any wooded area will do. The guidelines are simple and flexible: find some trees, whether a deep forest or a city park, and walk or sit or even meditate for a half hour or so. Depending on where you live, there are a multitude of types of conifers to visit and enjoy.

I have seen redwood trees in the California preserve named after John Muir, hiked a Nevada mountain to touch an ancient bristlecone pine, dodged chiggers in longleaf pines of South Carolina, and searched for pine nuts under a humble pinyon pine in southern Utah. Conifers are an ancient group of evergreen trees and shrubs with leaves reduced to needles or scales that produce seed-bearing cones— they are often resinous, and their pollen is distinctive and wind-blown. Conifers have been in the world for over 250 million years and have a worldwide distribution, where they grace many a national park or reserve; despite this, about one-third of the 630 species are considered threatened or vulnerable. We give them names to recognize their age or height: consider the bristlecone pine named Methuselah that clocks in at over 4,800 years old; the Senator, a pond cypress from Florida that lived to about 3,500 years old until a woman decided to smoke in a cavity in the tree; Old Tjikko, a Norway spruce in Sweden that is a 9,550-year-old clone; Gran Abuelo, a Patagonian cypress in Chile that is about 3,600 years old; the Mediterranean cypress called Sarv-e Abarkuh growing in Iran for perhaps 4,000 years; and the largest tree of all, General Sherman, a giant sequoia in California that is 275 feet tall and over 36 feet in diameter at the base. These record-holding trees are mighty, and their history is ancient, but they are vulnerable to global challenges including overharvesting, insect pests, climate change, tourism and trampling, conversion to agriculture, cattle grazing, and fire. Many are from fire-driven systems and require fire to sprout, but human interference and road construction has led to buildups of fuel under the trees, making fires more intense, and climate change has added to this vulnerability, as evidenced by devastating fires in Australia, California, and the Pacific Northwest as I was writing this book. Among the hundreds of species of conifer, several groups provide us with fragrant products or a place for reflection and peace. The list that follows is a sampling of some of coniferous species to whet your appetite for more, whether you visit them outdoors in the woods or indoors with a field guide. Perhaps we can take a moment and think of a favorite spot outdoors under some trees or next to a peaceful body of water and do a little virtual forest-bathing.[6]

Pines

Perhaps it is their age, their location, or simply the sound of the wind through the branches but ancient bristlecone pines (*Pinus lon-*

gaeva) also give us the gift of peace. Although they grow throughout the western United States in the arid mountains of California and Nevada and are fairly common, it is the old and gnarled trees that we treasure and remember. They are generally viewed after a hike through boulders and scrub and, in the case of the Nevada population, form a grove nestled below a splendid peak carved by glaciers looming up from the surrounding desert at Great Basin National Park. Bristlecones age slowly, sacrificing parts of themselves and keeping other parts alive so that many old trees look like mostly dead wood with a thin strip of bark keeping the tree alive. By the time they die the wood is hard and polished, like the stones of its surroundings. In addition to bristlecones, there are about one hundred species of pine throughout the Northern Hemisphere that can be found in a variety of habitats, generally with periodic fires that sweep through. As we have seen, they are hugely important economic species, mainly for wood and paper but also for resins and edible seeds. Retsina is a drink made with resin from the Aleppo pine (*P. halepensis*), and yummy pine nuts from the pinyon pine (*P. edulis*) were indispensable to early Native Americans but also, from other species, to lovers of pesto on their pasta. From petite bonsai to immense sequoia and fast-growing slash pine to ancient bristlecone, pines are found throughout the world and exhibit a variety of shapes. Just standing in a pine forest seems to reach all the senses with the wind in the boughs, fresh scent in the nose and lungs, rough and sticky bark, and the majestic view of a gnarled trunk bending to age or wind.

Fir

In southern Mexico the oyamel fir (*Abies religiosa*) grows at a specific altitude in the mountains, creating groves with the ideal microclimate to support millions of migratory monarch butterflies (*Danaus plexippus*) that winter in the trees. Optimal humidity keeps the butterflies from drying out, and temperatures are just right to allow them to hang from the trees in dense clusters without expending energy to keep warm. Mexican authorities and conservation groups are working to preserve these forests and the fragile butterflies while also allowing traditional harvesting activities. One project aids local women who prune the trees to gather boughs for Christmas wreaths and sell them to supplement their income. Balsam firs (*A. balsamea*) are found throughout northern North America, where they are popular with humans for building construction and Christmas trees and with wildlife for food and shelter. Moose, deer, squirrels, and other small mammals shelter within the vegetation; birds nest in the branches and eat insects from the buds; mammals, especially moose, and birds such as spruce grouse and ruffed grouse eat buds and branch ends. Resin dries to a clear, transparent film and is used for mounting microscope slides. For fragrance work there is an exquisite solvent-extracted absolute that is dark green and thick, highly aromatic, and sweet with a fragrance that many people refer to as jammy. If you are lucky enough to get a whiff, think the very best of evergreen aromas highlighted with open-air freshness, maybe with a bit of dark-under-the-trees-where-a-moose-slept, hints of resiny vanilla, and a feeling that it might be very good indeed dribbled on top of some raspberry jam and excellent bread.

Cedar

True cedars include trees valued through the centuries for their fragrant wood and stately appearance. There are just two species of true cedar, Himalayan or deodar cedar (*Cedrus deodara*), growing in the western Himalayas, where it is the grandest tree in the forest, and Lebanon or Mediterranean cedar (*C. libani*), originating in areas around the Mediterranean. Himalayan cedar is also called *devadaru* in Hindi, which means tree of the gods, perhaps deriving from the incenselike fragrance of the wood. Chips and sawdust are distilled to yield a sharp and elegant essential oil, like a typical cedar, but with a bit of sweetness and camphor. Lebanon has long valued its majestic cedars, which have been used for timber since the time of the Egyptians; wood from the trees built the Temple of Solomon. Extensive harvesting led the Roman emperor Hadrian to establish boundaries to protect the trees, protections that the government of Lebanon is attempting to emulate today.[7] Atlas cedars are a subspecies growing in the mountains of Morocco and Algeria that yield a beautiful essential oil, similar to that of Himalayan cedar. In Japan *Chamaecyparis obtusa,* called false cypress or hinoki, has been used for bathtubs and produces a lovely light and woody essential oil that is also just a bit sharply green.

Cypress

In between the upland pine forests of the southeastern United States, there are the black-water rivers that support forested wetlands. The water is brown with tannins, acidic, so full of insect and aquatic life that it is difficult to sleep for the noise of the frogs and humming of mosquitos if you happen to be camping on a river sandbar. With huge buttressed trunks and "knees" poking out of the water above the roots, bald cypress (*Taxodium distichum* var. *distichum*) trees shelter birds and provide structure to the forest. Studies have recently found bald cypress trees living in the forested wetlands of North Carolina that are upwards of two thousand years old. Pond cypresses (*T. distichum* var. *imbricarium*) are a variety that live in shallower waters that dry up occasionally.

Araucariaceae

Kauri trees (*Agathis australis*) are native to New Zealand, and their immensely tall, straight trunks were perfect for constructing canoes for the native Māoris, their extruded gum was used for fires, and the soot for tattooing. These massive trees are protected in New Zealand within a few forest preserves but are victims of an infection called collar rot and rarely exist outside of protected areas. Tāne Mahuta, or God of the Forest, and Te Matua Ngahere, or Father of the Forest, are tourist attractions in the largest remaining kauri stand in New Zealand. In Australia, wollemi pines (*Wollemia nobilis*) were discovered, or rediscovered, in 1994 in the Wollemi National Park of New South Wales and are recognized as living fossils, representatives of a species that were in the world more than sixty-five million years ago. The government of Australia quickly limited visitation and keeps the location a secret to protect trees from trampling and human-transmitted disease and also sends seeds out to botanical gardens around the world in an active effort to grow trees and conserve the species. As Australia burned in late 2019 and early 2020, the newspapers reported a concerted and so far successful effort by firefighters to save these trees in their secret groves.

Juniper

Often called cedar, both European and North American junipers in the genus *Juniperus* possess fragrant wood used in making pencils and storing clothing due to the durable nature and insect-repelling qualities of their woods. Juniper berries are used as a flavoring for gin and produce a nicely sharp and dry essential oil that I like to blend with citrus peel oils to give a bit of bite to the sweet.

Arborvitae

Thuja occidentalis is native to eastern Canada and the United States and was given the name by a group of explorers into Canada who were suffering from scurvy. Drinking a tea from the foliage of the tree is supposed to have cured the group, so they called it the Tree of Life. In Japan, the native arborvitae *Thuja standishii* is one of several trees used to build Shinto shrines and is currently grown in plantations for its useful, lightly scented wood. There are essential oils from arborvitae, but they may contain thujone, also a component of wormwood, which is an ingredient in absinthe and is found in sage. Caution is advised when ingesting or using thujone-containing products.

Yew

Yew trees (*Taxus* sp.) belong to one of the oldest taxa of conifers, and some specimens are truly ancient. One specimen found in Saint Cynog's churchyard in the small Welsh town of Defynnog has perhaps been growing for five thousand years. Yews produce a deadly toxin called taxine that featured in a detective novel by Agatha Christie called *A Pocket Full of Rye* in which a murder was committed with the poison: the murderer disguised the bitter flavor in English marmalade made of the bitter peels of Seville oranges. Christie worked as an apothecary's assistant and later as a pharmaceutical dispenser during both world wars and was familiar with medicines and poisons of the early twentieth century. The flexible and lasting wood of the yew was used to make longbows for archers of England and medieval Europe, giving them the name yeomen.

Sequoia

Sequoia trees are classified into just two species. Redwoods (*Sequoia sempervirens*) grow along coastal California, where they depend on incoming fogs for moisture and may reach record-breaking sizes of up to 367 feet in height with a 30-foot diameter at the base. A denizen of fire-driven ecosystems, redwoods are adept at resprouting after a fire, including from fire-killed stumps (and logged stumps), giving them characteristic burls that form on the trunks. They reach their massive trunks out of valleys and into the sky, creating parklike groves worthy of the dinosaurs and were so large when discovered by American loggers that it took five men three weeks to fell a single tree. People came to see the giants and waltzed on a dance floor made from a trunk of a felled tree or drove their automobiles through the hollowed-out trunk of a standing tree. The state of California was granted rights to the trees early on, and President Teddy Roosevelt, inspired and urged on by naturalist John Muir, signed legislation to protect the area where they were found. *Sequoiadendron giganteum* is the giant sequoia and grows in central California in the Sierra Nevada: it is equally large and impressive and may take up to one hundred years to reach maturity. Wildfires in central California during 2020 and 2021 have been so extreme that they have killed unknown numbers of giant sequoias as drought and extreme fires of climate change overcome millennia of adaptation. In 2021 at Sequoia National Park, park service personnel and firefighters wrapped the trunks of some sequoias estimated to be among the oldest and tallest trees in the world. General Sherman, at 275 feet and as much as 2,700 years old, is among them.

Podocarpus

Podocarpus or yellowwood trees are found mainly in the Southern Hemisphere and find their homes on many of the islands of the Pacific. Next to pine, this is the genus with the most species in the conifer group. Unlike pines, which may form uniform stands in dry soils, yellowwoods are often found in moist or wet forests scattered among other types of trees. Instead of tough, woody cones like pines, they produce soft, berrylike cones to attract birds and mammals that assist in dispersal of seeds, and they grow long, narrow leaves rather than needles.

Now, go out and find your local conifers, take a hike, indulge in a spot of forest-bathing, then come home and crush a bit of mint in your favorite beverage. Enjoy!

Bearded iris (Iris germanica) *flower and bud*

10

Perfume Notes

Before the fragrance industry perfected the art of getting aromatic chemicals from pine trees, the city of Grasse in southern France was a center of production for fragrant ingredients. Indeed, it still is. The complexity of extracting and using aromatics from plants meant that, for most of history, perfumes were reserved for the very wealthy, often royalty, who kept their own alchemists to distill and create fragrance from flowers, woods, spices, herbs, and musk. Therefore, perfumes smelled of flowers, woods, spices, herbs, and musk but most often flowers and musk, in varying ratios depending on whether you were queen or dandy. There were also perfumed gloves, heavily scented with—you guessed it—spices, flowers, woods, and musk to cover the raw scent of tanned leather. In the limestone mountains near the blue Mediterranean and surrounding Grasse, lavender was found, jasmine thrived, and glovers used fragrant products to process leather and make scented gloves for wealthy and royal customers, including Catherine de' Medici. As it served such prominent customers the industry grew in importance and established the Society of Glovers in 1724. Glovers may have used a formula similar to an old one for fragranced leather, called Peau d'Espagne, or Spanish skin, that combines rose oil, neroli, sandalwood, lavender, verbena, bergamot, cloves, cinnamon, and gum benzoin in a blend in which to steep the leather. A mortar and pestle is then used to grind civet and musk with gum tragacanth to make a paste with any leftover liquid from steeping the leather. The resulting paste can be put between two pieces of steeped leather and pressed until dry, giving the leather a lasting fragrance.[1]

Before too long, perfume-making replaced glove-making, and companies used aromatics from the surrounding mountains and florals that grew well in local fields to found dynasties. Large flower farms were established and factories built to supply French perfume manufacturers: some of the same companies are still in existence today. Grasse became a center of perfumery, with its crops of jasmine, violet, and tuberose plus a certain je ne sais quoi required to make and market iconic perfumes. The city is now a center of three stages in the making of a perfume—cultivation, processing the plants, and composing the perfume—but it is also a representative of an irreplaceable ecosystem. Skills and knowledge present there were recognized by UNESCO in November 2018 as the Pays de Grasse was added to the Representative List of the Intangible Cultural Heritage of Humanity. To quote UNESCO, "The practice involves a wide range of communities and groups, brought together under the *Association du Patrimoine Vivant du Pays de Grasse* (Living Heritage Association of the Region of Grasse). Since at least the sixteenth century, the practices of growing and processing perfume plants and creating fragrant blends have been developed in Pays de Grasse, in a craft industry long dominated by leather tanning." The announcement goes on to recognize the imagination, memory, and creativity needed, in addition to technical skills, to achieve this honor.[2]

Perfumers often work using three accords that comprise a top, heart, and base blend, a formula that gives structure and interest to a perfume. Citrus notes appeal because of their freshness and simplicity; they also tend to be short-lived and are perfect for the top notes that introduce a perfume. Other possible top notes may include spices like black pepper or cardamom and herbs like coriander and tarragon. Sensual beauty is found in the heart of the perfume, with florals like jasmine, neroli, and rose that unfold, sometimes with delicacy and sometimes with a wallop, after the citrus notes waft away. Woods, resins, and musks contribute to the long-lived base notes, in French called the *fond,* meaning background or substance, and are indispensable to the construction of a perfume. Like a perfumer I will construct the next part of the aromatic story as I would a perfume, beginning with the sparkling top notes of citrus.

If deep and luscious florals are the beauty of perfume and musky notes are the beast, citrus notes are the tiny fragrant bows that tie up a perfume for the perfect presentation. A spritz of citrus top notes is the effortless and happy greeting that a perfume grants you when you spray it. Sweet orange, mandarin, bergamot, grapefruit, and lime are at their very best when they can waft forward on a scented spritz, gifting you with the opening notes of your favorite perfume. With scents composed mainly of terpenes, different citrus species are characterized by a recognizable fresh citrus aroma, but it is trace amounts of more subtle aromatics that give each a uniquely different fragrance. For example, pomelo (a large, grapefruitlike citrus), orange, and tangerine peel oils share similar constituents, up to 97 percent of them terpenes such as limonene and citral, which have lemony fragrances. The remaining 2–3 percent may be up to forty minor constituents that create each different fruit's characteristic fragrance and taste, much like a perfume that is graced by unexpected accent notes. Essential oils from the citrus family have traditionally been cold-pressed from the peel—think of the spray of fragrant oil when you peel a fresh orange—meaning that they remain unaltered by the heat of distillation. Anciently the fruits were scraped by hand and the oil was gathered in sponges, and more recently oil has been separated from peel with the aid of mechanized rasps. Now citrus peel oils are often a byproduct of the juice industry to be separated from pressed juices, keeping the price low and acting as a value-added product in the citrus industry.[3]

When driving around Florida one can often see row upon row of short, dark green citrus trees in groves with, depending on the time of year, oranges dotting the foliage of the commercial operation. A few years ago, I was surprised to see these same green trees growing wild and dotted with bright orange fruits when I was hiking in a natural area thick with palm trees and live oaks dripping with Spanish moss near Lake Okeechobee. The trees were sour orange trees, likely *Citrus* × *aurantium,* offspring of dooryard citrus grown in Florida that were originally planted as rootstock for more fragile types of sweet oranges. Many times, the sweet orange trunk and branches of the tree would wither and die, but the sour orange roots would sprout and flourish, making a new tree and putting out flowers and fruit to seed themselves through not only Florida but into other parts of the Southeast. Homeowners are sometimes surprised when their orange tree stops producing sweet oranges and begins putting out small, inedible fruits, most likely the result of a dieback of the sweet orange tree and regrowth of the sour from the roots. Perhaps some will enjoy the beautifully fragrant flowers and learn to make marmalade from the peel, since the fruits are inedible. Oil from the sour orange is used to make such liqueurs as Grand Marnier and Curaçao, and some are grown to produce commercial essential oils from fruit, leaf, and flower. Sour oranges, sometimes called bitter oranges, are the source of orange flower essential oil called neroli, named after the seventeenth-century princess Marie Anne of Nerola, who reportedly used it to perfume her gloves and bath. The fragrance of the oil perfectly captures floral, citrus, and green blended with freshness and power that is beautiful from the bottle but does not quite capture the seduction of a sour orange tree in full bloom like my neighbor's when I picked blooms from a tree full of them on a warm spring dawn. Colognes would not be colognes without neroli, especially the

one called 4711, one of the originals and a two-hundred-year-old formulation from the city of Cologne, Germany. In addition to the flowers, leaves, stem tips, and tiny fruits may be distilled to produce
petitgrain oil, *petitgrain* meaning tiny fruit. A lovely oil, it is similar
enough to the more expensive neroli that it may be used as an adulter-
ant or on its own to provide some pretty woody notes, too.[4]

Bergamot (*Citrus × bergamia*) is grown strictly for the peel along
the southern Calabrian coast of Italy where terroir and weather make
the best fruits.[5] Possibly the descendant of a sour orange pollinated by
a lemon or something similar, the fruits are large and wrinkled with a
lemon shape and are basically inedible, but they produce a beautiful
and invaluable oil for the fragrance industry. Bergamot essential oil
(from the peel) is complex, beginning with a green sharpness to the
nose that gets your attention but quickly moderates to something sort
of citrusy but mostly deeply and freshly floral, a bit like orange blossom but more as if the fruit itself is blooming and opening. Bergamot
is a historic and important oil in the creation of colognes and perfumes, and it pairs nicely with lavender as they play off each other
with sweet, tart, herbal, and floral all at once. Put them in a bottle
together, add some light florals such as neroli with a touch of jasmine
over a base of frankincense-laced woods, and you have a beautiful
cologne. Bergamot should be used in body products, including perfume, with care due to the phototoxic nature of a constituent called
bergapten, a furanocoumarin that can cause dermatitis when applied
to the skin before going out in the sun. Berloque dermatitis is the
condition resulting from repeated exposure to even dilute amounts of
these chemicals, but the furanocoumarins can be removed from an
essential oil for skin safety. Lime peel oils may also cause dermatitis,
and bartenders making margaritas in sunny climates need to be careful they do not develop margarita dermatitis if they hand-squeeze
limes or lemons for the refreshing drink.

Let us go back for a moment and look at the evolution and ancestors of modern-day citrus. It is a complex search involving the pomelo (*Citrus maxima*), a sort of giant, sweet grapefruit with a very thick peel, and the mandarin (*C. reticulata*), but then, as in domesticated roses, family lines get complicated and lost in the clouds of time. There was probably a fair amount of asexual propagation, rootstock grafting, selection of sports or mutations, and breeding back and forth for desirable traits. Pomelos (also spelled pumelo and pummelo) came to us from Southeast Asia.[6] If you ever see a giant grapefruit-looking fruit but larger and somewhat flattened, it is likely a pomelo and worth the effort to get to the center of this fruit. Maybe just once, or you might get hooked. The peel is over half an inch thick, and I have yet to find a good technique to separate fruit from peel, but sectioning the fruit and biting out the sweet, tangy interior seems to work just fine. Keep a bit of the peel to enjoy the smell, too. Unlike the large pomelo, the other root of this family tree, mandarins, are small. Domesticated varieties are sweet and easily peeled. Mandarin peel oil is usually cold-pressed for the essential oil and, like the fruit, can be sweet and fresh but must be used within a few months or it loses the friendly character to become a bit sharp and bitter. Some oils will have a bluish fluorescence when diluted with alcohol due to the presence of N-methyl-anthranilate, a compound that can have a floral grape-soda fragrance but may also, at least to some people, smell a bit musty.

Citron (*Citrus medica*) is a fragrant type of citrus that may have originated in India in the foothills of the Himalayas and is likely the golden or Persian apple referred to by Theophrastus. An ancient lineage, it is thought to be one of the ancestors of modern cultivated types, giving rise along with pummelo and mandarin to form three groups: limes and lemons, grapefruits, and sweet and sour oranges. The Buddhist and Hindu god Kubera was shown holding a citron, and the Hindu god Ganesh is also associated with the fruit. Called

Etrog in Hebrew, citron is used in the Jewish Feast of the Tabernacles. One of my favorites, Buddha's hand fruit, is a form of citron, also fragrant, and having many fingers reminiscent of the hand of the Buddha. The peel is the best part, the interior being basically inedible, and one or two of the fruits can be kept in a bowl to fragrance a room with an aroma that is mostly floral with just the lightest of citrus.

If you look closely at the peel of an orange or hold a lime leaf up to the sun, you can see tiny vesicles full of fragrant oils. You may also see, as with my little lime tree, tiny caterpillars looking very much like bird droppings that are the young of the giant swallowtail (*Papilio cresphontes*). My small lime tree in a pot in the front yard is seemingly miles from other citrus of any kind, and yet each year graceful female butterflies find it and flit over the leaves, doing quick dips to deposit single eggs on the new and fragrant leaves. She, like many other butterflies, has followed the fragrant plume of scent from the lime tree to find the perfect host for her young. As they grow, the caterpillars remain a mottled brown and white but also develop a large thorax that resembles a snake's head, albeit a tiny snake's head. But that is not all. If you poke these little somewhat dangerous-looking critters, they immediately evert a brightly colored gland called an osmeterium from their head that resembles the tongue of a snake and secretes a strong fragrance composed of various terpenes. These terpenes come from the leaves of the tree in the caterpillar's diet and are also excreted in the feces, or frass, of the caterpillar, as I found out when I raised a few in a butterfly enclosure. A strong and earthy blend of citrus with a touch of what I can only describe as caterpillar musk would waft out as the creatures deposited their abundant leavings in the bottom of the enclosure.

When I open a tiny bottle of jasmine grandiflorum in my perfume classes, even diluted to 10 percent, it is mere seconds before the fragrance fills the air. It is full-bodied, lush, warm, and somewhat overwhelming, unapologetically floral with very subtle earthy, even fecal, accent notes, and a bit of honey. The more adventurous students, often those familiar with natural ingredients, will enjoy the effect of the full-bodied aroma. Those who are not may create a nice perfume, but perhaps with a heart that lacks dimension or feels slightly flat; these students can usually be persuaded to add a bit of jasmine. Even below the level of perception, jasmine makes all the difference in the world, adding a roundness or fullness to a perfume. Jasmine absolute, the solvent-extracted ingredient, is one of the top three perfume ingredients, and the French have been known to say, "No perfume without jasmine." I agree.[7]

There are about two hundred species of jasmine native in tropical areas of the Old World, and most have white flowers or white with hints of pink or yellow. Many have long, tubular corollas for pollination by moths, and many also have dark and fleshy fruits for dispersal by birds. Lovely flowers and pretty green foliage have made them popular garden plants, and they can be found everywhere as cultivars. As with so many scented and precious plants, Arabs brought jasmine to Spain during the Moorish occupation and planted it in their gardens. The English have also taken to the plant, and many gardens feature jasmine. Jasmine has inspired poets and artists, perfumers, and tea makers, and for perfumers it is the ultimate white flower. Flowers of *Jasminum grandiflorum* cannot be distilled to produce an essential oil for perfumers, but their habit of continuing to breathe out their fragrance lends them beautifully to the process of enfleurage, or they can also be solvent extracted to yield an absolute, the perfumer's more affordable choice. Jasmine grandi, as it is often called, may also be referred to as Spanish or common jasmine and is the species grown in Grasse for perfumery. Sometimes it is called *Jasminum officinale,* a slightly different type and a different species, that may be called poet's jasmine. For perfumery, it was one of the key ingredients in early perfumes, including Joy by Jean Patou and Chanel N°5 by Chanel—the original formula of which was reported to have 4 percent jasmine absolute. Although I reach for jasmine grandi most often for its rich, indulgent fragrance, I also appreciate the quirky green floral of *J. sambac*, which also has hints of rubber. When I teach a male student, I often have him take a sniff of jasmine sambac absolute, which for some reason is often thought of as a masculine floral and does appeal to men. For a more unique extract, we have the indolic white flower scent of *J. auriculatum* and the slightly spicy and fresh absolute of *J. flexile. Jasminum auriculatum* is grown in India, and the absolute is available in smaller quantities than either grandi or sambac but can be a valued addition to a perfumer's palette. Although they share many constituents, each species has a unique blend with, for example, higher levels of indole, which adds dark floral-fecal tones to *J. auriculatum,* or the fresh and minty methyl salicylate in *J. flexile. Jasminum grandiflorum* has an abundance of sweet floral notes, and *J. sambac* has more green and fruity notes.[8]

Jasmine sambac is the source of the white flowers that form head-dresses for brides in India and the flowers in jasmine tea. The double form, called Grand Duke of Tuscany, is probably one of the most beautiful flowers I have ever seen—it looks like a tiny, creamy rose with hundreds of perfect petals immaculately arranged—and smells of white flowers with hints of fruit, green stems, tiny bits of wood, and fresh, indulgent jasmine. Sambac is the national flower of the Philippines, where it is known as *sampaguita* and made into garlands and crowns. In Hawai'i jasmine sambac is called *pikake,* and the single-flowered variety is used for leis. Buds are picked in the morning once they turn from green to white and are strung soon thereafter; more than eighty buds are required to make a single lei strand of about thirty-six inches, and the fancier the lei, the more flowers needed. Javanese use flowers in traditional ceremonies, including weddings, and gather three kinds of flower—jasmine, rose, and cananga (a relative of ylang-ylang)—in a collection called a telon. The group of three is composed of flowers with three colors and three aromas. White from jasmine sambac represents holiness and the fragrance a clean and soft heart, red from rose indicates strength and the fragrance a bold and honest attitude, and yellow from cananga stands for simplicity and the fragrance humility. Cananga may refer to ylang-ylang, whose yellow flowers are leggy with long petals that emit a soft and elegant, slightly spicy, and truly tropical fragrance. This is another tree that must be enjoyed on a humid morning in the heat of a tropical summer as the octopus-looking flowers crowd the branches and lengthen around a green center.[9]

Jasmine flowers, as mentioned, must be harvested in the morning well before noon, and they must be picked by skilled workers who are able to avoid damaging the flowers, since damage causes release of indole. Indole will affect the scent, making it more funky than perhaps desirable and may result in pinkish-brown spots on the white petals. While Grasse is thought to produce the highest-quality jasmine, most of today's jasmine comes from India and Egypt, where the plants are grown under contract to produce the extract and provide a living to many local families. As with rose, the expense of the absolute comes mainly from labor, with about eight million flowers required to yield just under five pounds of the concrete that is extracted to get just over two pounds of the absolute. Hugely expensive as a natural ingredient, a jasmine effect is often obtained by synthetic bases that might include some of the flower's constituents, including methyl jasmonate (one of the jasmonates discussed in the context of tobacco flowers earlier), benzyl acetate, milky fruity lactones, and indole. However, even a small amount of real jasmine extract is indispensable to attaining a true jasmine effect.[10]

In an effort to get out of my head recently, I took a clay hand-building class at a local make-it-yourself studio and decided to work with a piece of flat-rolled clay that the instructor had imprinted with a pretty piece of lace. The pattern was a lovely floral, and I had a fun time working on a cup, or was it a pot? Maybe a pencil holder? Once fired, it was a bit off-kilter, but the lace pattern came through. I found some glaze in a periwinkle color and applied the first coat, then a second. The floral pattern from the lace seemed to disappear in the uniform color of the glaze. I tried an overlay of neutral beige, which sort of helped. This being do-it-yourself, the instructor was absent at this point, or she might have pointed out the fine-tipped brush and a dark brown or green glaze that I could have used to outline or highlight the flowers. Had I been trained in classical painting techniques, admittedly a bit of overkill for my first clunky piece in clay, I would have perhaps used an underlay in dark neutral colors to give interest to periwinkle flowers. Classically trained artists used layers of paint, beginning with dark warm colors or even neutral gray, that gave depth and contrast to the rich colors in the final layers. These dark colors were carefully laid down, muting the bright white gesso of the first layer, and would show through as shadows or create darker midtones below the final, top layer of pigment. This darkness worked with the bright and light much as salt enhances sweetness in cooking and gave compelling form to finished paintings. Flemish painter Daniel Seghers was known for his paintings of flowers in the sixteenth century, and as I looked through his work, I saw gorgeous flowering jasmines, peonies, tulips, and irises made luminous against a dark background with the play of light and dark in the details of their petals. This is how I think of indole—the dark that brings out the light and that makes the floral of jasmine rich and resonant and interesting.

Among the complexity of constituents that contribute to the fragrance of jasmine flowers, three interact to give the floral, the fresh, the rich, and the creamy: jasmonates, methyl salicylate, and lactones. These are made more interesting by the addition of tiny amounts of indole. Jasmines produce volatiles called jasmonates in their flowers as part of their white flower fragrance that induces moths to approach for nectar and pollination. If you are a botanist studying plant defenses or plant-insect interactions, you have heard of methyl jasmonate and jasmonic acid. If you are a perfumer, you have definitely heard of a derivative of methyl jasmonate called Hedione. It was fragrance chemists who first isolated and described these molecules and botanists who later found their reason for being. Methyl jasmonate was discovered in 1957 by a fragrance researcher named Édouard Demole, commissioned to find the missing "something" in jasmine constituents at a time when only a few were known. Jasmine absolute production was limited and very expensive, and yet perfumers included at least a bit of jasmine in about 80 percent of perfumes. Methyl jasmonate, isolated from Egyptian *Jasminum grandiflorum,* turned out to have a very "jasmine" fragrance: specifically, it achieved with a single molecule both a certain floral character and the richness of jasmine extracts. Methyl jasmonate had a fatty, buttery, deeply floral fragrance that perfumers found to be exquisite and evocative of jasmine flowers. A slightly different form, methyl dihydrojasmonate, was later obtained by hydrogenating (reacting with hydrogen) methyl jasmonate. This altered form was, on first sniff, very lightly scented and more subtle but was christened Hedione after the Greek word *hedone* for pleasure and promoted to perfumers instead of the more expensive methyl jasmonate. Samples were sent out to prominent perfume houses for their perfumers to evaluate and, it was hoped, appreciate. It took a few years, but perfumer Edmond Roudnitska created Eau Sauvage for Dior with 2 percent Hedione for a jasmine floral heart, blended with citrus, woods, and herbal notes for a new kind of masculine perfume. This was the breakthrough that the molecule needed, and today it is appreciated for its synergistic and elegant effects, even at small doses, adding a subtle power, notes of soft lemon, freshness, diffusiveness, and bloom to many types of perfume. Various forms of methyl jasmonate are now produced and marketed by fragrance companies to different effects in flavors and fragrances."

Nearly twenty years later, botanists began to study the action of jasmonates, both methyl jasmonate and jasmonic acid, in plants, and they have since found them to function in protection against chewing and sucking herbivores but also as communication compounds. As we saw with tobacco, jasmonates are produced and emitted in response to wounding by herbivores, and they do two things. First, they elicit what is called a direct response for protection by stimulating the plant to produce protective compounds such as the neurotoxin nicotine in tobacco plants, and second, they communicate with predators and parasites of the herbivores. There is also a third action: nearby plants can eavesdrop on the fragrance and respond to the danger that is afoot, or a-wing, without being damaged themselves. In some plants, rather than putting the chemicals into the air, jasmonates may work within the plant to stimulate the secretion of extra floral nectar, nectar found outside the flower, to attract predatory arthropods such as ants to protect delicate petals against herbivores. This means that jasmonates act in both inter- and intraplant communication, and somehow the plant can distinguish between signaling for production of defensive chemicals and signaling for nectar secretion. Jasmonates also help to defend the plant against necrotrophic pathogens that can cause plant tissues to die.[12]

Another white flower ingredient that may be produced and released by plants for protection is methyl salicylate and its derivatives. While the jasmonates protect against herbivores and necrotrophic pathogens that may be introduced into plant tissue by chewing and sucking insects, salicylic acid protects against a different kind of pathogen, one that produces disease rather than causing tissue death. Scientists have found what they call crosstalk between jasmonate and salicylic acid in plants—both are inducible defenses (activated upon harmful stimulus), and each inhibits production of the other in a sort of reciprocal antagonism. Which means that if the plant emphasizes production of jasmonate over methyl salicylate, it may protect against chewing insects but be vulnerable to invasion by disease-causing pathogens and vice versa.[13] Methyl salicylate has a sweet, root beer, candy-type, fresh fragrance often also described as minty from its association with candies. In willow trees it is found in the bark as salicylic acid and is found in wintergreen plants, while in tuberose, stephanotis, and frangipani flowers it is part of their fresh white-flower fragrance designed specifically to be attractive to moths. In the voodoo lily (*Amorphophallus konjac*), methyl salicylate triggers heat production, causing the flower to warm rapidly and release its stinky fragrance to attract fly pollinators. Methyl salicylate is a benzenoid, one of a family of floral scents that includes vanillin, eugenol, and others. In humans it may provide pain relief, used anciently direct from the plant, and more recently the related acetylsalicylic acid is used as aspirin. Various forms of the chemical are blood thinners, and overuse of aspirin or muscle ache creams containing wintergreen oil or a combination of the two may cause an overdose and death.

In Mexico where it originated, this next white flower is called *omixochitl,* or bone flower (*Polianthes tuberosa*). In French, the name is *tubéreuse,* and they are one of the famous flowers of Grasse. In English we call them tuberoses. Sometimes pink or red, they are in the agave family and have been cultivated in Mexico since before the Spanish came to the New World. Tuberoses are one of fifteen related species growing mainly in forests, where they are pollinated by hawk moths. One of the best things about tuberoses is how the fragrance changes between day and night, even in cut flowers. Daytime fragrance is fresh and green, with a diffusive sweetness that is just a bit rich, while at night the scent becomes more languorous, heady, and sensual, exhibiting that white-flower fragrance so attractive to moths. Flowers are generally harvested early in the morning, when the fresh scent is more dominant, and are currently extracted with solvent to produce a gorgeous concrete and lovely absolute, but tuberoses were also used for enfleurage in the early days of Grasse.[14]

Two other plants were important in the history of Grasse: iris and violet are beautiful in your garden and beautiful in the heart of your perfume, but it is not often the flowers that provide the scent. Violets (*Viola* spp.) have a history of surprising me; at my former home in northern New York State, they peeked out from the edges of the boggy lawn in spring, and for a while in my Florida yard, a scattering of tiny purple flowers would reseed in various pots and occasionally do well enough to bloom. Shrinking violet is a misnomer: they are unabashed in seeking out pollinators. Violet flowers will change position to display their colored patterns and to attain the best orientation for contact between pollinator and floral genitalia. Pretty white and yellow patterns on the lip help to guide a pollinator into the center of the flower, ensuring that reproduction will occur. First come the heart-shaped leaves that may peek up through late snows and give rise to flowers that appear early in the spring so they can take advantage of any pollinators that are out and about. Then they show off their colorful throats with nectar guides that may be decorated with raised hairs the bee or fly must wiggle through to find the flower's interior and complete the exchange of gametes for sexual reproduction. But the more frequent mode may be asexual by way of flowers that grow low on the plant and produce seeds without fertilization to be ejected in a process called ballistic dispersal to sprout near the plant. Alternatively, seeds may be carried away by ants attracted by the eliasome, or fatty body on the seed, to end up on ant-created waste heaps that are often rich in nutrition and loose soil. Despite the higher success rate of asexually produced flowers, violets continue to produce lovely flowers in shades of purple and white that are lightly scented. They are found in temperate regions around the world and there are about five hundred species, but since they hybridize readily this is just an estimate. Some of the most endangered species are called metallophytes, meaning metal-loving, and they grow on soils impregnated with toxic heavy metals, usually lead or zinc, in Europe around old mines. Two types of *Viola lutea,* or zinc violet, from Germany and northern Europe grow in soils contaminated by atmospheric deposits from old metal smelters—habitat that is hazardous for humans.[15]

Viola odorata is the sweet violet grown in Grasse, the south of France generally, and Italy. These violets were a favorite of Empress Josephine, who had them embroidered on her wedding dress. The flowers can be used in salads, to make sweet syrups, or for decoration, and the leaves thicken soups. When it was available, violet essential oil was difficult to produce and overly expensive; even now it is not produced for use in perfumery, but there is an absolute of the leaves. The leaf absolute is difficult to appreciate, being fecund, darkly leafy, and maybe a bit as if you put your face right down atop rich soil and crushed an abundance of violet leaves in front of your nose along with a few flowers. However, at very high dilution, the green fragrance becomes less overwhelming and the delicate floral beauty peeks through to give you both flower and leaf. Expense and popularity of violet fragrances led to an urgent desire to determine the constituents of violet essential oil to enable synthesis of replacement molecules. In 1893 two scientists, Ferdinand Tiemann and Paul Krüger, began the search and determined that they could work with orris root from irises rather than violets because it had a similar scent profile and a higher percentage of essential oil, making it more economical. Unfortunately, the molecule they isolated did not exhibit the fragrance of violets. Disappointed, they cleaned the lab and washed the glassware, which at that time meant using sulfuric acid. Fortunately, they caught a whiff of violets coming from the cleaned glassware as the iris chemicals interacted with the acid. This was the real stuff, and they named it ionone, which we now know consists of an alpha and a beta form that together give violets their sweet, diffusive, and slightly woody fragrance. Both α-ionone and β-ionone are popular perfumery ingredients still, long after they were used to create the first modern violet perfume, Vera Violetta by Roger et Gallet, in 1892.[16]

Ionone is also present in one of my favorite plants, tea olive (*Osmanthus fragrans*), an unobtrusive plant with a gorgeous smell that I finally discovered with the help of my friend the botanist. I could smell the sweet fragrance every winter in South Carolina but could never find the flower. Then she showed me a humble-looking bush and the tiny white flowers that produced the big scent. Tea olives have a type of flower that seems to have a stronger fragrance away from the plant, and search as I could, I never would have found the plants without the help of a professional. Winters are cool and dry in the South, and the fragrance of tea olive seems at its best in the freshness that allows the apricot, fresh floral, leathery, tealike beauty of the tiny flower bloom. Osmanthus flowers are, as the name suggests, used to scent tea, and the solvent-extracted absolute is gorgeous for a leathery white flower effect. Native to China, where they are one of the ten most famous flowers, Osmanthus flowers may range from ivory to green-white to deep orange, and it is the carotenoid in the color that is altered by the plant to produce its lovely α- and β-ionones.[17]

Irises are not related to violets taxonomically, but as Tiemann and Krüger sensed, they are related by their scented compounds. The commercial fragrance of irises is not in the flowers, although some are scented, but in the roots. Iris roots are rhizomatous, meaning thick and fibrous, anchoring the plant, and spreading out to produce new shoots. Harvested and aged, they yield a powder called orris that has been used for millennia to scent body care products, incense, and linens and is today distilled for orris butter, which has a beautiful and ethereal powdery violet fragrance. Orris roots are harvested from *Iris pallida* from the Florence, Italy, area and *I. germanica* in Morocco. Roots must be dried for up to five years, then ground and treated with dilute sulfuric acid, at which point they can be steam distilled to yield a thick product with varying amounts of different irones (chemically related to ionones), each type contributing to the scent. Orris root oil is today one of the most expensive perfume ingredients available but can be used in tiny amounts to achieve a powdery and floral effect.

Irises are named for Iris, goddess of rainbows, and have flowers of many colors, sometimes a beard, and occasionally a fragrance. An iris flower's architecture is designed to guide pollinators toward their pollen, and they accomplish this by providing a landing platform and a runway or nectar guide of a different color and even sometimes, if it is a bearded iris, by adding texture just like the much smaller violet flower. Plants may invest in floral color patterns and shapes differently depending on pollinator: the copper iris (*Iris fulva*), a wild species in Louisiana, produces red flowers with reflexed petals to attract and accommodate hummingbird pollinators. Hummingbirds sip nectar while on the wing and need the flowers to be bowed back to make room, and if the anthers protrude from the flower, as they do in copper irises, then the bird is more likely to pick up pollen. Since hummingbirds lack a sense of smell, the red-flowered iris does not need to invest in fragrant constituents. Bees, on the other hand, respond both to the strong patterns of nectar guides and to fragrance, and the zigzag iris (*I. brevicaulis*) attracts bees with its blue and white flowers, yellow nectar guides, and a strong floral scent. Dixie iris (*I. hexagona*) has large purple flowers and yellow nectar guides to bring in bumblebee pollinators that fit the large flowers. These flowers illustrate a multistep process where pollinators may recognize color and pattern from a distance and respond by approaching, at which point nectar guides and fragrance will further pull in the bees and bumblebees.[18]

Now we come to the beast at the base of many perfumes, described by a simple four-letter word that refers to a diverse and complicated set of fragrant ingredients: musk. Animalic, sensual, earthy, leathery, spicy, fecal, floral, and transforming—the descriptors range from stink to sublime. Persistent musk notes linger to blend with the skin, give the perfume an intangible and complex depth, and add lift to a floral heart. Musk ingredients are generally animalic in origin and are long-chain, slowly evaporating molecules used for communication. They say, "I am here, this is my place!" or "I am here, and I am available!" spreading the word regarding territory or sexual availability. Some-times the two purposes overlap. Traditional musk ingredients have come from mammals including beaver, musk deer, and civet, and har-vesting the animal's musk has involved captivity, death, and mistreat-ment. Musk aromatics are also rare, expensive, highly regulated in some cases, and not a tool most perfumers use today since they can substitute synthetic replacements. Here is the thing about natural musk: like any wild animal, there are ways to approach it. For many people, the world is a more interesting place if they know there is wildness in it, perhaps a little danger, hidden deep in a fragrant jungle or pacing on the grassy savannah. Animalic musk is the slightly dan-gerous ingredient in perfumes through history, but it works best when dilute. If you were to stick your nose into a full-strength bottle of musk, your nose and your brain would not know what to do with it—it overwhelms and intimidates. But dilute it down, way down, put that danger a safe distance away, and it has facets and layers that often produce a different perception for different people. Musk molecules are large and complex and do not evaporate quickly, which is why they are perfect base notes—they last and last on the skin to anchor the more ephemeral notes or fix a perfume. But they also do wonders for the floral heart of a perfume, the beast to the beauty, showcasing the floral, whether it is rose or jasmine or neroli.

For the male musk deer (*Moschus* spp.), producing musk means that the scent he deposits in his environment will last and last, informing local females of his location and availability and local males that they should keep away. Musk deer are small deer that live in dense vegetation in mountain valleys throughout much of Asia. These deer do not have antlers, but both male and female have large front tusks. Musk deer are solitary, crepuscular animals and use scent rather than vision to delineate shared spaces by way of communal latrines, where droppings and urine are deposited near their daytime beds and at the boundaries of their territories. Marking during the rut by males is completely different: a small, specialized pouch on the belly contains musk-scented grains to produce the scent. Trade in musk once took place from Tibet, recognized as a source of fine musk, along the Silk Road trading network. Musk deer are endangered everywhere they are found, and nearly all trade is prohibited under international law as regulated by CITES, the Convention on International Trade in Endangered Species of Wild Fauna and Flora. All species are endangered, not only from hunting to provide musk for the perfume trade and for traditional medicine, but also from habitat loss as humans encroach on their range for livestock use and cut down trees for agriculture. Although only adult males bear the pods, musk hunters and trappers may kill deer indiscriminately: for every pod, at least two deer and probably more are killed. There have been attempts to breed deer for pods, which can be removed without killing the deer, but these have largely been unsuccessful in part due to the animal's solitary nature. Some countries are working to preserve habitat and educate locals in an effort to sustain local musk deer populations. What does deer musk smell like? Once it is diluted in alcohol, it has been described as animalic, the very definition of musk, slightly sweet, possibly a tiny bit floral, evocative of skin, and very persistent. In perfumes it will add lift and has the effect of smoothing the notes and helping them work together—it does wonders for florals, but the animalic note should not be noticeable.[19]

Beavers (*Castor canadensis*) produce castoreum, another animalic musk marker, which they deposit on their dams and lodges and even scent mounds built out of pond mud that create a sort of scented fence to mark territory. Given all the work that beavers must do to cut trees, build dams, dig out and fortify lodges, it only makes sense that they will mark their efforts and keep others away. Castoreum is produced in anal glands and stored in paired sacs near the anus. Occasional stories in the news refer to the use of castoreum as a flavor ingredient, usually in ice cream. The stories are true. Castoreum is listed as Generally Recognized as Safe, or GRAS, by the U.S. government for use in strawberry- or vanilla-type flavors. But the extract, generally an alcoholic tincture, is also described as leathery and sweet, great for perfumes with a leather theme.

Totally not appropriate for food and almost never used in perfumery today, civet is also leathery, quite animalic, and urinous in tone. Once an important perfume ingredient, its disuse results from the potential for animal cruelty in obtaining the secretion. Civet cats are not really cats but long-bodied nocturnal mammals from tropical Asia and Africa and comprise two groups, true civets and palm civets. Most produce a musky substance from their perineal gland, but true civets of the subfamily Viverrinae are known for their particularly aromatic secretion. Civet paste is traditionally obtained by scraping either the glands on the animal or deposits from a marking post in their territory, but animals may be killed or kept in cages to obtain it.[20] Because of concerns about animal use for perfume ingredients as well as expense, the perfume industry has discontinued using civet along with nearly all animalic musks and now uses synthetic musk chemicals. For civet there is a synthetic version called civettone, musky, sweet, dry, and diffusive, that may work on big cats (the real thing) as well as humans. Zookeepers have known for a while that big cats such as ocelots, jaguars, and cheetahs love perfume, particularly Calvin Klein's Obsession for Men, and will roll and rub their faces in it and other types of perfumes that are applied to objects in their enclosures. Anyone who has had a domestic male cat may know that cats will mark their territory, and big cats do this, too. Apparently, this tendency also stimulates an interest in musky perfume ingredients, and zookeepers have been known to request perfume donations to use for big cat enrichment. Researchers have also explored using perfume as a lure to draw cats into camera and hair traps, where they either trigger a camera to take a photo or leave a bit of hair behind for DNA analysis.

Ambergris, or gray amber, is one of the few noncommunicative musks, is expelled by the whales that create it, and was once the darling of dandies and queens. Today, it is rarely used in its pure form. Instead, chemists have spent decades isolating and refining the molecules that provide the long-lived musky notes of this by-product of whales. Ambergris is released into the ocean from the digestive tract of a small percentage of sperm whales (*Physeter macrocephalus*) and is mostly composed of a waxy-fatty substance that the whale produces to coat the sharp beaks of cuttlefish—the whale's primary prey item— as they move through the digestive tract. This wax, together with intestinal waste, may build up in the digestive system of about 1 percent of whales to form an indigestible ball called a coprolith that is eventually expelled as a smelly black mass, sometimes weighing as much as two hundred pounds. A large part of this waxy mass is a compound called ambrien. It is unclear whether this activity is normal defecation or whether the waxy blob grows in the intestines of a small number of whales until it become a potentially fatal blockage. Possibly both.

Once released, ambergris gains its magic by floating on top of the sea exposed to the salty brine and sea air, sometimes for years (some samples have been found to be more than a thousand years old), to land on a beach somewhere as jetsam ambergris. What started out as dark and fecal transforms into gray or even gold and silver with aromatic notes that vary depending on the sample with the fragrance of sea, salt, ocean air, tobacco, moss, incense, seaweed, flowers, wine, and musk. There are a few tiny pieces carefully wrapped in cloth and stored in a wooden box in my ingredients cabinet—the fragrance rarely reminds me of the beach but smells of rich moss or sweet dirt, salty and floral breezes, and a fecal note that is elusive and not entirely unpleasant. The origin of ambergris was long a mystery, and various theories existed: perhaps it was the excrement of sea birds that lived on sweet herbs by the ocean or a blend of beeswax and honey from shore-living bees; it might be truffles from the bottom of the sea or the drool of dragons that sleep on rocks by the sea. Perfumers might call it ambra, and one must be careful not to confuse ambergris with amber, the fossilized resin of pines. The waxy lumps are neither solvent extracted nor distilled but should be tinctured in alcohol to a low concentration of 3–5 percent or so. As with so many aromatics, this opens it up and allows the molecules to arrive at the nose in somewhat distinct little packets to reveal all its complex notes. As a flavor it may be added to tobacco or liqueurs to round and mellow.[21]

There are two more animal musk products that do not require killing, cutting, or scraping—hyraceum and beeswax. The small mammal called the rock hyrax (*Procavia capensis*) shelters in rocky outcrops, or kopjes, of South Africa and gathers vegetation into its den, where it cements the plant material with urine and feces, creating a highly aromatic mass called hyraceum or amberat. Hyrax dens may be inhabited for many years—some go back generations—preserving the hyraceum, which yields a record of plant material within the gathering range of the small mammal over time. What also happens is a hardened, funky and smelly glob of local plant and animal material that ages within sheltered dens to become less fecal or urinous and more earthy, haylike, leathery, and musky, described by some as a conglomeration that evokes the smell of South Africa and the small mammal that makes its home there. Pack rats do the same thing in North America, as do other rodents around the world, but it is the hyrax of South Africa that produces a fragrance worthy of perfume. Beeswax absolute comes from unprocessed beeswax that is melted, strained, and then, most likely, washed with alcohol. I include this among the musks because of the fragrance, which I have found to be sweet and honeyed but perhaps less than expected: it is agrestic and somewhat pheromonal and yet useful for smooth tones in a perfume. As an intimate constituent of the homes of bees who often communicate by smell and where fragrant honey is stored, the wax, I like to think, is full of a honeyed scent.

Very few plants produce true musk notes, but perfumers will use musk ambrette, patchouli, currant bud absolute, and oakmoss for tenacity and a certain musky interest. Seeds from ambrette or musk okra (*Abelmoschus moschatus*) yield a beautiful example of vegetal musk, with a soft and elegant presence that suggests musk with hints of powder and skin. Tiny ambrette seeds nestle in the pods of the tall, prickly-leaved plant, which is related to hibiscus and okra. Flowers are typical of hibiscus in pale yellow with a deep maroon or brown center that only last a day. Edible when soft, the pod dries out to produce many kidney-shaped seeds that are extremely hard but hint at ethereal aromas hidden inside. The seeds may be distilled to produce an essential oil with a high percentage of fatty acids that need to be separated out before it is useful. Ambrette comes from India and can be cultivated in tropical regions, including my little Florida backyard, where the plants once grew to five feet tall and for several months produced gorgeous yellow flowers that faded by the end of each day to be replaced by the next bloom. Green fuzzy pods followed that, once dry, could be opened to extract the tiny seeds. Whether or not the conditions of my yard and my methodology were correct, it was fun growing the beautiful plants and getting a hint of ambrette from the tinctured seeds.

Oakmoss (*Evernia prunastri*) is a reindeer lichen found in southern Europe, where it grows primarily on oak trees. Lichens are a partnership between moss and fungi, with each partner providing something the other needs and becoming, sometimes, more than the sum of their parts. Oakmoss is an old perfume ingredient and is harvested to produce an absolute that is earthy, inky, often darkly green, leathery, metallic, or even seaweed-y, and once was indispensable in creating a class of perfume known as chypre. In the base of a perfume, it adds tenacity and naturalness along with a certain dark interest, standing up well to the spices in a variety of perfume types and providing contrast to the green and floral of the others. Oakmoss and related tree mosses are now banned for use in perfumery except in tiny amounts due to their potential for skin irritation, and many of the older perfumes have had to be reformulated with various substitutes. Currant buds (*Ribes nigrum*) when extracted with a solvent produce an absolute that is complex, not at all fruity, and most often equated with cat pee. This is another one that requires high dilution and tiny amounts in a blend to give it something perhaps animalic, perhaps dirty, but definitely interesting. My sniffer detects fruity and winey notes along with a deep, deep green that is almost not-green and animal that is not really animal but plant pretending to be animal: it is a very distinctive ingredient.

Patchouli is what perfumers call a "love it or hate it" fragrant ingredient and yet has been a natural ingredient of major importance to the perfume industry. Patchouli (*Pogostemon cablin*) is an herb in the mint family that grows in tropical Asia but is cultivated in a variety of locales. The plant has been used in traditional medicine in the East and came into fashion in Europe in the mid-nineteenth century, when the leaves were wrapped with imported luxury scarves to keep moths away. Queen Victoria often wore a knitted shawl scented with patchouli and is responsible for this early fashion-fragrance mashup. Patchouli is forever famous for its association with the hippies of the 1960s as an earthy natural perfume useful to cover up the scent of cannabis. It is not possible to describe the scent of patchouli without using the words earthy or dirty, but one needs to also add the descriptors wine, raisins, tea, musk, sweetness, and depth, and it is well worth searching out a beautifully done distillation with all those aspects. Historically, and even today to some extent, the leaves were locally distilled in iron stills that give the oil its dark color: much of the patchouli essential oil available today may be brown-amber unless specifically processed to remove the brown. Patchouli essential oil acts as a fixative, thanks to large molecules in the oil, and does wonderous things to roses, evoking their connection to earth and highlighting their floral beauty. Sometimes when I am out and about, I will get a whiff of fragrance from a passerby, and the immediate impression is that it is a nice perfume, but then my scent memory will kick in and I realize the wearer just has really good taste in patchouli. Often when I work with patchouli, I dab my wrists with droplets from the bottom of a pipette (it is one of the few essential oils that is okay to use full strength on the skin), and the wool rugs and pretty scarves stored in my closet also benefit from the scent of patchouli dropped on a cotton sheet to wrap the rugs and diffuse into the basket holding my scarves. It makes them smell pretty and repels insects at the same time.

Perhaps it is fitting that I finish this section with patchouli, one of my favorite essential oils. From the humble plant comes a fragrance that is ancient, that is iconic to both queens and hippies, and that has a complexity of scent that defies accurate description. Plants are masters at blending fragrance, and they do it for a variety of purposes, adapting to need and environment. As humans will do, we have taken those fragrant products and tried to make them our own. Perfumers have made the blending and use of aromatic constituents created by plants into science and industry.

FRAGRANCE AND FASHION

By the mid to late nineteenth century, perfumers and companies had the science, the bottles, and the demand—a demand they seem to have both discovered and encouraged—all the moving parts that allowed them to complete the move from perfume as medicine to perfume as a statement of fashion. It is here that scented molecules themselves become the story. Although this is a story about the *natural* history of fragrance, I hope, dear reader, that you have figured out that molecules are the letters and words (and possibly paragraphs) of the story. The first synthetic aroma molecule was created in 1866, inspiring and enabling perfumers to transition from a small and specialized art to a commercial industry that reached out to a wider audience. Synthetic aroma molecules are made in the lab, and the process can be controlled to yield a pure product with a replicable scent and affordable price, unlike essential oils, which have complex blends of molecules and an intrinsic variation because plants grow in different places and in diverse terroirs. Just as important, these new scented molecules also inspired perfumers to create products unlike those found in nature: they were concepts and abstracts rather than florals and musks. Now perfumers could build perfume types like white flowers: just as a flower mixes various molecules to produce a characteristic scent that will attract pollinators, so perfumers blend aromatics to achieve their vision of a white flower and attract the human nose.

Industrialization implies the mass production of perfumes, ingredients being churned out of factories, and prices being gauged for purchase by the many. But first perfumers began experimenting with the effect of synthetic molecules to create what has been called a fantasy perfume—something not found in the natural world but an abstract concept of a fragrance. The first modern fantasy perfume began with the smell of vanilla, clover, tonka beans, and sweet hay from a molecule called coumarin, synthesized in the lab in the 1860s. From there, researchers moved on to create synthetic vanilla and molecules with the fragrance of violets and jasmine. The fragrance industry today is built primarily on such molecules, often derived from petroleum but also from natural starter materials such as turpentine and even from yeast. Generally inexpensive and consistent in fragrance, these are the backbone of the modern fragrance industry.

Lilac (Syringa vulgaris) *flowers*

11

Impossible Flowers and Building a Perfume

By the twentieth century there was the science of perfumery, at least as far as combining the various constituents for creating a blend that evoked, for example, that elusive white flower aspect. Never mind that flowers have been doing it for millennia; scientists and perfumers could now make a list of molecules derived from an evolutionary process that produced a world of flowers capable of making scented compounds. Many of these molecules are common across varied types, colors, and shapes of flowers. The scent of a flower is not simply floral but a combination of smells that blend to make, for example, a lilac smell different from a violet. Aromatic chemicals also act as modifiers to add freshness, earthiness, and herbal, minty, or even fecal and musky overtones to the floral. Some of these molecules have a low odor threshold and can be sensed at extremely minute concentrations, even as low as ten parts per billion. They are called high-impact molecules and can impart a desirable or unique character to a blend.

Jasmine, an iconic white flower, is extracted via solvent due to the huge expense of enfleurage, and neroli can be both distilled and extracted by solvent. No extract of gardenia is remotely affordable, being mainly an artisan product of enfleurage. Tuberose and frangipani extracts are available but costly. These flowers fall into the loose category of white flowers, with blooms that open at night when they can attract moths with molecules including linalool, benzenoids, alcohols, and esters along with other common volatile organic compounds. But the flowers smell noticeably different to the extent that most of us could likely tell the difference blindfolded because white flowers are possibly the best perfumers of the plant world. Like a natural perfumer, they use some of the most common and lovely of the fragrant molecules to create their aroma. Linalool is a molecule with a characteristic floral fragrance that is found in a variety of plants and attracts diverse groups of pollinators. Despite seemingly different odor profiles, coriander, rosewood, lavender, and basil all contain linalool as part of their unique blend of aromatic molecules. If you smell a true lavender essential oil, you can experience the simple and calming sweetness of linalool—to my nose it is freshly floral with hints of green backed by a bit of sharpness. When extracted from basil, the linalool is accented with an herbal fragrance, and from rosewood it is beautifully floral-woody. Terpenes, such as limonene, add a lemony or citrusy fragrance to a floral scent and are integral to the uplifting scent of citrus peel and flower. Jasmine flowers, rich and indulgent, have a touch of indole, which is often described as earthy or fecal. What sounds like an unpleasant smell adds a depth or richness to a fragrance mix: much the same as a dash of fish sauce adds umami to a complex dish. Methyl salicylate gives the characteristic fresh and minty smell to wintergreen but is also a contributor to the fragrance of tropical blooms like ylang-ylang and tuberose. Benzaldehyde (sharp and sweet cherry-almond) and benzyl alcohol (rosy and balsamic) add to white flower fragrances and appeal to moths as well. Eugenol for a hint of cinnamon spice and rosy geraniol may also make up the perfume of a white flower. As with citrus, each plant may add its own twist to the white floral fragrance, whether the jasmine-y jasmonates for jasmine, the mushroomy hints of gardenia, the green undertones in tuberose, or the smell of leather in tea olives. Perhaps there is a garden nearby where you can sniff a couple of different white flowers to see if you can sense the fresh in a daytime tuberose, the green in citrus flowers, or the mushroom in a gardenia.

Most perfumers and perfume houses have formulations for different types of floral concepts, including those for which it is impossible, exceedingly difficult, or very costly to extract the scent. Over and over I am told, when people find out I am a perfumer, that they love the scent of lilacs. And could I make a lilac perfume? Unfortunately, a true natural extraction of lilac is exceedingly rare and expensive. The same is true of the flowers of peonies, sweet pea, lily of the valley, tobacco flowers, and violets, to name a few. It may be that the yield is exceptionally low, requiring many flowers for tiny amounts of the extract, as with violets, or the flowers just do not lend themselves to extraction. So what is a perfumer to do? As with white flowers, many will use a blend of extracts, aroma chemicals, or isolates to achieve the same fragrance profile. For example, the magnolias in my backyard have a beautiful fresh floral fragrance with hints of lemon. I found that a blend of the gorgeous Australian flower boronia (*Boronia megastigma*), some jasmine, and a hint of lemon comes pretty close to the scent, to which I add a base of citrusy Australian sandalwood and a touch of vetiver.

Lilac fragrance is built from various components, including what are called lilac aldehydes, that are also used for other floral fragrances. These aldehydes, also called syringa aldehydes, are derived from linalool and were described from lilacs but are also found in various other flowers, wines, and fruits including papayas and plums, and they are attractive to such pollinators as small moths and mosquitos. White campion (*Silene latifolia*) has been the model plant to study lilac aldehydes because it produces them to attract its moth pollinator, a noctuid moth (*Hadena* sp.). The moth uses the plant to obtain nectar but also as a host plant for its larvae: the female lays her eggs on female flowers so that the larvae can feed on developing seeds. As with other white-flowering plants, white campion has a strong floral scent that is attractive to nocturnal pollinators. One study in Switzerland

and Spain found that male flowers emitted more lilac aldehydes, one of the known compounds that elicit antennal and/or behavioral responses by the moth. Male plants also produced smaller but more strongly scented and abundant blooms to enhance the waft of lilac perfume into the night air. Male moths were more attracted to the highly scented male blooms than to less fragrant female flowers. Female moths were a bit less choosy, visiting both male and female flowers without significant preference. What happens in the moth's antennae, you ask? Scientists have developed a system to tie the response of a moth's antennae to a specific scent using both gas chromatography and what is called an electroantennographic detector, or EAD. When a particular scent is wafted over the antennae of the moth, the EAD indicates the strength of the response. In one study, antennae of the moths responded to several scented molecules but reacted in a linear fashion to both low and high concentrations of the lilac aldehydes, basically sensing them in both low and high amounts, indicating particular sensitivity to those compounds. Other compounds had a threshold of odor strength, meaning that a certain strength had to be present before it elicited a response from the moth. Behavioral observations in wind tunnels where different scents were wafted toward the moths also indicated a preference for the lilac aldehyde fragrance.[1]

Blunt-leaved bog orchids (*Platanthera obtusata*) are found in wet areas across most northern parts of North America, Asia, and Europe, where their green flowers and low height allow them to blend in with the surrounding vegetation. To attract pollinators, they emit a slightly green, musky scent that brings in a variety of moths but also those denizens of bogs: mosquitoes. One study placed trained participants in an orchid-rich bog to observe as mosquitoes (*Aedes* spp.) entered the tiny flowers in search of nectar, triggering an internal spur to attach pollinia to the head, often the eye, of the mosquito for it to transfer to the next orchid flower. Observers, using visual counts as well as GoPro cameras, counted fifty-seven nectar feeding events (there is no record of people-feeding events) as mosquitoes quickly followed the scent to find flowers, land, and probe for nectar—getting pollinia attached to their eyes in the process. Mosquitoes are attracted by two compounds, lilac aldehyde, which has a floral lilac scent, and nonanal (an aldehyde with a waxy green floral scent), that are produced by bog orchids. *Platanthera obtusata* produces a fragrance that has higher amounts of nonanal compared with lilac aldehyde, and it was found that the antennae of *Aedes* mosquitoes responded with specific electrical activity when exposed to that particular ratio between lilac aldehyde and nonanal—what may be called a Goldilocks blend that has to be just right. Other orchids with higher amounts of lilac aldehyde and less nonanal were not as attractive to *Aedes* mosquitoes but perhaps, as we saw with campion flowers, were more attractive to small moths and other pollinators.[2] Perfumers use lilac aldehydes for green and floral impact and to add sweetness as they build the impossible fragrance of lilacs or to make their hyacinth and rose blends more complex and interesting.

Muguet is the French name for lily of the valley (*Convallaria majalis*), a small springtime plant treasured for its delicate aroma. I know that my sister and I checked the cool, shady north side of my mother's house to track the emergence of the pure white delicate, bell-shaped flowers to bring them inside for their simple presence and delightful scent. This beauty hides a darker side. All parts of the plant carry a load of cardiac glycosides and are poisonous: perhaps that is why the deer left the lilies of the valley alone to grow and multiply when they ate all the other pretty flowers in my mother's garden. It is impossible to extract the flowers for perfume use—perhaps it is just as well—and there are formulas for creating the aroma of this impossible flower. Soliflore is a term for the perfumer's vision of the scent of a flower, like lily of the valley or lilac. This class of perfume may be based on the floral extract itself, as was originally true with Vera Violetta by Roger et Gallet, but may also be achieved by skilled blending of individual ingredients, including single molecules.

In aid of scent creation, recipes or receipts exist from as early as those written on the walls of Egyptian perfume houses. Housewives, druggists, soap makers, likely even the manufacturers of patent medicines, all had formulas for various blends and ingredients for flavor, scent, and nostrums. Formulations were published for the homemaker in various flyers and books on the cosmetic arts. George William Septimus Piesse listed fewer than one hundred natural fragrance ingredients for his recipes for useful products such as soaps, creams, and pomades, smelling salts, hair dyes, and absorbent powders.[3] My mother was raised during the Great Depression and kept many of the frugal habits learned during that time, including making her own lotion that she used throughout her life. More than any perfume, I remember the distinctive scent of her witch hazel, glycerin-based blend: even now I struggle to describe the fragrance other than lightly herbal with a hint of citrusy-sugary sweetness.

As more and more natural ingredients with various origins and scent profiles become available, the perfume wheel becomes more complex, but it continues to be a handy way to group, compare, and contrast. For example, among the natural ingredients you might have groups such as herbal versus green, with lavender and fennel coming under the herbal class and green describing more intense fragrances such as violet leaf or galbanum. There is the dry freshness of juniper berry or the sweet spiciness of cinnamon. Woods may be sharp or buttery, sweet or earthy, and florals range from heavy to soft and powdery to green and sharp. There are fruity citruses and fruity florals, such as tea olive and boronia. Vendors may group their molecules and plant extracts according to such classifications. On one website listing only natural materials I counted more than four hundred options; many were variations according to source country and location or were chemotypes, as we saw with rosemary—cineole versus verbenone, for example. There were absolutes and concretes, attars, carbon dioxide extracts, resins, and waxes. Now imagine the large number of potentially synthesized single aromatic molecules. Sometimes a single molecule such as linalool or vanillin can be extracted from the source: linalool from lavender, vanillin from vanilla beans. Those molecules would be called isolates and present a simpler fragrance than lavender or vanilla and would, perhaps, be more useful for some purposes. There are a variety of other processes to create through chemical reactions many of the molecules in use today. Depending on the process and the source material they may be considered natural or synthetic: regulations defining these categories are complex and differ between the European Union and the United States. As a thought exercise, consider the number of constituents in lavender essential oil or in pine that may number in the hundreds, and then multiply by the number of fragrant plants in your world.

Hay-scented fern (Dennstaedtia punctilobula) *frond and fiddleheads.*

12

Scented Worlds: Industry and Fashion

We have arrived at this final chapter on perfumery as well as the book as we move into all-new territory. The last half of the nineteenth century and up to the time between the world wars saw changes in European cities to improve sanitation through the construction of water and sewer infrastructure. As the stink of uncontained sewage and the smells of unwashed bodies began to be reduced, perhaps it was not so necessary to use heavy musks and spices in fragrances as coverups. Homes of the wealthy began to be constructed with indoor plumbing, allowing people to wash more frequently with indulgent products such as perfumed soaps.[1] Bad smells became unacceptable, and perfumes became concepts rather than a reflection of the natural world, while fragrances moved from heavy musk and ambergris to such lighter scents as florals. Advances in chemistry and manufacturing as well as interest in fashion following World War II led to further changes in the perfume business. It was about this time that the first heavy-duty granular detergent for cleaning clothes was introduced. Tide was an immediate success with its light fragrance and efficiency: the brand continued to follow, and inspire, fragrance fashions, setting standards for the smell of "clean."

But the story of the original synthesized aromatic chemical, coumarin, begins before people were writing down formulas and before there were little amber bottles to hold medicine. It begins when people gathered medicine from their surroundings. As it has been through history, plants that have a pleasing smell were thought to be good for body and soul. Coumarin is the sweet in sweetgrass, the freshness of new-growing hay, honey in sweet woodruff, and vanilla in tonka beans but also appears in many plants, fungi, and bacteria. Sweetgrass (*Anthoxanthum nitens*) grows in North America and northern Eurasia. Long used as an herbal medicine, on the Great Plains sweetgrass was brewed into teas and made into salves and infusions. It was also important as a spiritual medicine, used to purify and cleanse the spirit: as with incense in the Old World, smudging with sweetgrass provided a visible sign that prayers were ascending. The sweet smell of this grass is apparent as you bruise it with your feet and only gets better when you cut and dry it for use as a basket, a smudge, or a medicine. Persistent and pleasant, it offers hope. Growing from underground rhizomes, sweetgrass spreads and fills in lowlands and moist areas with welcome green in the spring. Sweet woodruff (*Galium odoratum*) lends its sweetness to German *Maiwein,* or May wine, and bison grass (also *Anthoxanthum nitens*) adds its herbal vanilla flavor to a Polish vodka called Żubrówka.[2]

Bison grass grows and European bison roam in an old-growth forest on the border between Poland and Belarus, protected as a UNESCO World Heritage site since 1979. The Białowieża Forest is ancient and complex and messy—it is a place where trees fall to remain where they lay while their dead wood nourishes complex food webs beginning with fungi and insects that in turn feed birds and small critters that may end up food for wolves and other carnivores. Giant trees such as the ancient pedunculate oak (*Quercus robur*) decay into the soil to nourish an abundance of other plants, including the local bison grass. In early 2016 the Polish government attempted to increase logging in the forest district, endangering the protected Białowieża Forest with its irreplaceable habitat. To create awareness of the beauty and value of the 350,605-acre forest as it was being threatened, a partnership between conservation groups and online gamers mapped the area in excruciating detail using satellite imagery, maps, and photos. That map was incorporated into the video game Minecraft to reach out to the millions of users in the community and visually illustrate the loss that logging would create. Did that help stop the logging that was taking place in and near the forest? Many people think so.

Before coumarin was discovered and isolated in 1820, a number of chemicals had been isolated from their plant sources, but none had been specifically created from an alternate source and later synthesized from unrelated individual molecules. Coumarin's spicy, balsamic, and sweet vanilla scent is valued by perfume and flavor industries, but it may also add a hint of tobacco to a perfume blend or enhance lavender in a scent. Coumarin was important in the creation of one of the first modern fragrances, Fougère Royale, by Houbigant (1882), what we might call a fantasy fragrance—something not found in nature.[3] In this example, a fougère type of perfume is inspired by the idea of what a fern would smell like (*fougère* is the French word for fern). Although some ferns have a light fragrance, the scent of a fougère perfume is an imaginary concept created by the perfumer using the unique scent of coumarin. Coumarin is one of a related family of chemicals used for blood-thinning medications, and food products containing it are regulated in the United States, including Żubrówka as originally formulated, and tonka beans as flavoring.

In addition to the fougère type, an effect achieved with oakmoss, lavender, and coumarin, perfumes may also be classed as chypre, spicy ambers, green, floral, gourmand, woods, and so on. Chypre-type perfumes get their name from the island of Cyprus and feature variations around a formula of mostly Mediterranean plants such as bergamot, oakmoss, and labdanum as well as patchouli. Vanillin was synthesized a few years after coumarin and used in the famous perfume Jicky, by

Guerlain (1889), which also contained coumarin. Jicky was another early ground-breaking perfume for its conscious vertical structure. The original formula had citrus top notes with a rose-jasmine heart and a base built on vetiver, orris, and patchouli, a touch of civet, and the vanilla tones of balsam, coumarin, and vanillin.

Musky ingredients have been important in the creation of scented products from the beginning almost of time. But as we saw earlier, many are directly from animals and are not often used any more due to expense, rarity, and perception of cruelty to the animal. Single-molecule musk aroma chemicals have been used as a replacement, beginning with the nitromusks, an accidental discovery. In 1888 an explosives chemist was seeking new kinds of explosives and found that one of his TNT-related experiments yielded a molecule (the nitromusk) with a pleasant musky smell. It was an immediate commercial success, and somewhat later, he developed a musk ketone that was also popular.[4] You may remember that musks are generally large molecules that persist in the environment as signals of sexual availability. They also persist on our laundry and in our perfume, accounting for their popularity, but some nitromusks have a similarly long life in aquatic environments, leading to concern for our rivers and streams. In the early 1900s nitromusk replacements, generally musk ketones, were developed, and there are a variety available, produced and marketed by different fragrance companies.

Where pricing and value of a perfume was once based on the cost of the materials, perfume companies began to realize profit in assigning value to an abstract concept such as desirability, and perfumes joined the world of fashion. Fashion designer Paul Poiret was perhaps the first to market a fragrance as a complement to a clothing line with the launch of a perfume house, Parfums de Rosine, in 1910. But it is the creator of the little black dress who is most remembered for combining perfume with fashion. Gabrielle "Coco" Chanel was a designer of fashions, famous for simple yet elegant clothing, the little black dress, and the perfume Chanel N°5, which forever connected fashion and fragrance. Her first driving principle for the perfume was that it should be unique and modern. As the chemical industry moved ahead in both isolating and creating fragrant molecules, the perfume industry followed suit to make the molecules themselves the centerpiece of scented products. The creation of one of the most famous perfumes, Chanel N°5, was not the first scent to rely on isolated molecules for effect, but it has become the iconic representation of the modern perfume. Coco Chanel was tired of women smelling like flowers and so requested a fantasy perfume of her own that would fit with the new spirit of the 1920s. She wanted a perfume that would make a woman smell like a woman, would reflect her preference for cleanliness and grace, and would sell to the forward-looking woman of the time. Chanel N°5 was born from this concept and the creative mind of perfumer Ernest Beaux. Flowers are nowhere in the name, and the classically austere bottle is quite different from the typical ornate perfume bottles of the time. Despite the abundance of expensive florals in the composition, the aldehydes—the unique and sparkling top notes—are what garnered attention and continue to do so. These waxy, green, floral, and fresh aldehydes with hints of citrus peel gave a unique lift to both florals in the heart and woods in the base. Called octanal, nonanal, and decanal, or aldehydes C8, C9, and C10, they are found in citrus oils, among other plants. Chanel N°5 made aldehydes famous and were one more early example of synthetic aromatics used in a perfume.[5]

François Coty, the "Emperor of Perfume," earned a fortune from his business of both perfumes and cosmetics at a time when people were beginning to pay attention to hygiene. Scented soaps, perfumed vinegars, powders, and creams were all part of the business, as were perfumes packaged in beautiful and unique perfume flasks made by

Lalique and Baccarat. American soldiers in France after the end of World War I bought luxury items to take home, bringing the business of perfume to the United States. One of Coty's best-selling items was a gift box with cologne, powder, soap, and cream with the customer's favorite scent, easily purchased as a gift for a loved one. His first perfume, La Rose Jacqueminot, encompassed his vision of packaging and sales that began his long career and presence in the perfume and fashion world.[6] Charles Frederick Worth, another entrepreneur, opened a design house in Paris in 1858 with partner Otto Bobergh that was to become the House of Worth. Worth was a dressmaker of note and appointed court designer for the Empress Eugénie of France who also created fashions for leading actresses and singers Sarah Bernhardt, Jenny Lind, Lillie Langtry, and Nellie Melba. It was Charles's grandson Jacques Worth who added perfume development to the company products with several beautiful perfumes packaged in Lalique bottles between 1924 and 1934. Rather than continue with this story, I will invite you to visit your nearest department store and peruse the perfume aisles to see the abundance of brands tied to fashion or celebrities. Or watch for perfume ads during your favorite television show.

Many of the top ten flavor and fragrance companies had their start in the mid to late nineteenth century by founders with a knowledge of and interest in herbal medicines, fragrance chemistry, and essential oils. Such companies were well poised for success when people were seeking new scented products, infrastructure was coming into place, and science was advancing. Grasse was moving from making gloves to making perfumes, and the laborious process of enfleurage was being replaced by solvent extraction. Grasse was also near ports and the fashionable city of Paris. With beginnings mostly in France and Europe, the original companies were expanding throughout the world by the twentieth century to place facilities in countries where botanical and other resources were located. Asia and the Middle East have more recently established companies of their own.

By the end of World War II the fragrance industry, like the rest of the world, was dependent on machinery and technology, synthetic chemicals, and mass-manufactured bottles and labels. Marketing and close ties with fashion brought the industry increased wealth and presence. By the latter half of the twentieth century, technology gave flavor and fragrance companies a new tool and marketing concept—headspace analysis using Solid Phase Microextraction, or SPME, techniques, in which tiny amounts of volatiles are captured within a vial and analyzed. Headspace analysis does not require cutting down or damaging a plant but allows for analysis of an individual flower or an iconic plant for replication in the lab. This is the same technology that scientists began using to understand fragrances in plants as science gave them new tools.

Customer preferences constantly change, and in the late twentieth and early twenty-first centuries, with consumers asking for natural fragrances and flavors, biofermentation technology has taken off. Petroleum and turpentine starter stocks have long been a primary source of aroma chemicals for flavors and fragrances that are generally regarded as synthetic. Recent consumer demand for natural ingredients and the innovation of new technology continue to drive exploration in these industries. For thousands of years people have used yeast to preserve their food, create alcoholic beverages, and make new flavors. Companies learned years ago how to use yeast to produce vitamins, and it turns out that the same process can be used to produce specific aroma compounds. Yeasts are single-celled fungi that digest their food externally—they grow within or on the medium that provides their food. This means that they take in various carbon sources for energy, digest them, and excrete them into the same medium, producing alcohol for fermented beverages, flavor in chocolate and coffee, and the rise in our bread. Out and about in the world, yeasts are naturally found on the skins of fruits such as grapes, apples, and peaches, in the nectar of some plants, and even floating about in our oceans.

Yeasts such as *Saccharomyces cerevisiae,* one of most cultivated and used living things on the planet, have been fellow travelers with humans as we have transported grapevines, coffee beans, and cacao nibs around the world. Wine was produced in the Middle East more than nine thousand years ago, creating specialized strains of *S. cerevisiae* that produced favorable fermentation products. A genetic study of *S. cerevisiae* yeasts from wine found that these various yeast strains are genetically similar and are distributed, together with grapevines, along paths of human migration. Yeast types from North American oak used to make casks as well as from certain soils around the world also contributed to the genetic makeup of these strains. It is a little more complicated to follow genetic origin and diversity of the yeasts that work their magic in coffee beans and cacao nibs. Cacao trees (*Theobroma cacao*), as we have seen, originated in South America, were cultivated in Central America, and were exported to be grown in Africa. When cacao nibs are fermented, the microbes responsible come from plants in the local environment as the pods are picked and handled. These yeasts and bacteria do their work over about five to seven days of fermentation to digest the pulp and create flavor and color in the nibs, making chocolate. Coffee (*Coffea* spp.) may be either dry- or wet-processed over a period of hours or days in a process that also involves local yeasts, bacteria, and other fungi. Yeast (*Saccharomyces cerevisiae*) populations from both coffee and cacao fermentation, though related to wine strains, show greater diversity and geographic differences: South American coffee and cacao are quite distinct from African cacao and coffee. These regional groups may act together with soil, geography, and weather to provide a microbial terroir that gives flavor to chocolate and coffee.[7]

Humans have used these little invisible factories called yeasts to alter how our food tastes, making cheeses, sourdough bread, kefir, and cured meats into new foods that could be kept through the winter or during food shortages. Pickles, soy sauce, miso, beer, wine, sauerkraut, vinegar, and sausages all require the actions of yeast and sometimes bacteria as well. Recently, there has been a revival of fermented tea called kombucha, made from fermented black or green tea using what has been called the Manchurian mushroom starter culture, or scoby (symbiotic culture of bacteria and yeast), for the fermentation biota. One of the things that happens in fermentation that is different from cooking or baking is a creation of new flavors specifically resulting from actions of yeast on existing ingredients. Soy sauce does not taste anything like soybeans, wine is a whole new kind of grape juice, and yeast puts the rise in bread. My grandmother's mustard pickles combined with my mother's roast beef were a Sunday treat and took cauliflower, cucumbers, and tiny onions steeped in vinegar to a new level of taste.

Fermentation has traditionally been accomplished with various microorganisms, for example in milk products, sauerkraut, and kimchee, but early in the twentieth century scientists began investigating the possibility of using pure microbial strains to produce specific molecules. Yeasts, sometimes with bacterial helpers, can take nutrients such as sugar, protein, or alcohol and convert them into different molecules including amino acids, esters, alcohol, and aromatic compounds. Some yeast species are specific in the aromatic molecules they create. For example, one type will produce fruity banana and pineapple aromas from such agricultural leftovers as the husks of coffee beans, or another the peachy scented γ-decalactone from castor oil. Yeast-produced flavors and fragrances are marketed as natural, and some, such as γ-decalactone, are approved for food by the U.S. Food and Drug Administration. The basics of production are the yeast, some sort of nutrient such as sugar or oil, a vat where the process takes place, and a way to isolate and purify the resulting aromatic molecules. At the end, the product may be a fruity flavor for your favorite beverage or the smell of rose for your hand lotion.[8]

There is an increasing trend toward the preference for natural in such products as food and cosmetics. One source of those natural flavors and fragrances, as we have seen, is from plants and animals by way of distillation or other extractions.[9] Growing plants takes space for fields, input from fertilizers and other chemicals, water, and fuel to run farm machinery, creating environmental costs even for organic production. Growing plants for fragrance and food (and traditional medicines) provides incomes for perhaps millions of households around the world and often supports ancient cultural practices, traditional medicines, and arts. However, demand for natural products is such that there simply are not, for example, enough vanilla pods in the world to support all the ice cream and gourmand fragrances we demand. Like yeast-produced aromatics, most of our vanilla flavor comes from a vat or a factory, and both the United States and Europe include microbially produced flavors as natural.

Fragrance and flavor are now injected into products for our everyday life from paper towels to bottled tea to pet food. One large company estimates that many people will interact with their products more than twenty times a day, and these companies are nothing if not intensely aware of consumer preferences. The late twentieth and early twenty-first centuries have seen the rise of emphasis on natural by world players in the industry but also from artisan and niche perfumers who may reach a smaller market but bring their passion to all aspects of their craft from earth to bottle. Every commercial fragrance website I have visited emphasizes its commitment to small farmers, green technology, sustainability, and/or social justice with, it is hoped, enough information for concerned consumers to ferret through the words to find an aesthetic that matches their ideals. And so, we have arrived at the science and technology of the fragrance world—taking apart the essence of flowers and plants to make a tool. The tools are beautiful and make our lives prettier, but they are tools nonetheless.

Whether using turpentine as starter material or programming yeast to produce specific molecules, both flavor and fragrance industries use the reliable and inexpensive aromachemicals produced this way. Although people have been creatively extracting and blending scents for millennia, we can point to the isolation and subsequent synthesis of single molecules as the beginning of modern perfumery, giving perfumers the ability to use molecules as tools for creation.

For most of us, wherever we live on the globe, it is hard to imagine a world where fragrances for the home or the body are not fashioned out of a process in which carefully blended molecules create scent from the imagination of the creator. And where they are equally attuned to the preferences of the consumer. We have taken the molecules from the plant, created them in a laboratory, and focused them for our own use. However, we have also seen a rise in the popularity of back-to-nature products, bringing consumers, in some ways, nearly full circle. We want natural, we like the idea of natural, and we want to support companies that thoughtfully use the earth's resources. At least that is what fragrance companies have been finding, researching, and investing in. This desire is also exemplified by the surge in popularity of aromatherapy and an interest in natural products. The past twenty years or so have seen resurgence in natural and artisan perfumery following the publication of Mandy Aftel's book *Essence and Alchemy*, an in-depth exploration of natural perfume ingredients, their history, and their uses for making scented luxuries. Some of the best stories about perfumery are experienced through applying one's nose to the curious and unusual in aroma, and her book is a guide to those unusual fragrant ingredients. Another book that allows you to explore smells is *Nose Dive: A Field Guide to the World's Smells,* by Harold McGee, which covers various fragrances from tiny fungi to the smells of the universe as it informs you of commonalities and differences between various molecular sources of smell. If you have ever wondered what makes the feet of a dog smell of tortilla chips, he has the answer. Or perhaps you wonder about the subtle differences between various truffles or what the universe might smell like. Again, he has the answer. As the visual world of device screens has proliferated, also in the past twenty years, and sight is now our primary sense, it is perhaps a time to train your nose. Outside is a good place to do this, in a garden or at a park or indoors in your own home or a local store: How does the smell of the kitchen differ from the laundry room? Explore your local grocery store or department store to see if you can find the cardboard smell of packaged goods, the green of freshly washed lettuces, or (and this may be an easy one) the perfume counter.[10]

People and plants together have scented stories that range from humble garden to the sweeping events of time. Through most of our history plants were medicine, fragrance was a force for goodness, and humans have venerated and valued aromatic plants. We have mashed them in oils to use as unguents, burned their woods and resins for religion, traded our gold and lives for their spicy seeds, traveled the world to seek out new types, honored our dead with flowers, tended those flowers in a variety of gardens, extracted their scent with heat and steam, found scented molecules that support industry. Explorers, entrepreneurs, royalty, gardeners, and scientists have sought the source and secret of scent. Those scents may bring us peace in the form of a pretty candle in times of stress, a lavender bath at the end of the day, a spritz of perfume to perk up our spirits, or a straggly bouquet from our garden. Plants, for their part, produce the molecules just for themselves—they are the architects of scented smoke and the breath of flowers like that iconic white flower, the gardenia, that seems to be laughing at us as it blows perfume abundantly into the air.

Glossary

3,5-dimethoxytoluene Also called DMT, a major scent compound in the tea fragrance of roses.

absolute An extraction from a concrete using alcohol.

angiosperm Flowering plants in which the seed is enclosed in an ovary.

anther The part of the stamen that bears the pollen.

anthesis The time during which a flower is fully open and receptive.

attar (1) Rose essential oil. (2) The product of floral distillation into sandalwood or another carrier.

base note The foundation of a perfume, often composed of woods and musk notes, that assists in the longevity of the perfume.

bee A flying insect with four wings that gathers pollen and nectar.

beetle An insect in the class Coleoptera that has hardened front wings.

benzenoid A class of aromatic hydrocarbons containing a benzene ring and found in floral scents.

benzyl acetate A constituent of floral scent with a sweet floral jasmine fragrance.

bumble bee A large social bee in the genus *Bombus* that pollinates both wildflowers and crop plants.

butterfly A diurnal flying insect in the order Lepidoptera that has brightly colored, scaled wings.

caffeine A bitter-tasting compound found in plants such as coffee and cacao that protects against herbivores and may limit the germination of nearby plants.

carotenoid A molecule in a plant responsible for yellow or orange colors.

carpel The female reproductive organ of a flower.

caryophyllene A fragrant molecule with a woody, spicy, clove scent. The α form has a woody smell, and the β form has a woody, spicy, dry clove scent.

chemotype A varietal difference found in essential oil plants with differing chemical makeup but the same morphology.

cineole A monoterpene also known as eucalyptol that has a medicinal, eucalyptus-like fragrance.

cinnamaldehyde A fragrant molecule with a sweet cinnamon, spicy fragrance.

citral A terpenoid with a sweet lemony scent.

concrete An extraction of aromatics, usually from a plant, using a solvent.

corm A rounded underground storage organ of plants.

corolla The petals on a flower.

coumarin A vanilla-smelling constituent of some plants.

crocin A colored pigment found in saffron.

damascones A group of aromatic compounds found in rose oil, giving the oil its characteristic scent. Also found in tea and tobacco.

distillation The process of separating components or liquids through boiling and condensation.

eugenol Also called oil of clove, a liquid phenol with a spicy clove scent found in flowers. Methyl eugenol is used by insects such as fruit flies as a pheromone.

euglossine bee Also called orchid bee, a glossy green or blue insect that collects fragrant compounds that the males use to make a perfume.

eusocial When referring to animals, a group with advanced social organization such as found in ants and honey bees.

fecal Relating to feces.

fermentation The change in an organic substance by the action of enzymes, often caused by microorganisms such as yeast.

fly A small insect in the order Diptera that has two wings.

fungus Organism that feeds on organic matter and reproduces by means of spores. Includes molds, yeast, and mushrooms.

furanocoumarin A constituent of some citrus fruits that may cause dermatitis on exposure to the sun.

gamete Single male or female cells that fuse during sexual reproduction.

GC/MS Gas chromatography and mass spectrometry, used to separate individual aromatic compounds in a blend along a column and then identify them based on mass.

genotype An individual's genetic makeup.

geophyte A plant with an underground storage organ or root, such as a bulb, tuber, corm, or rhizome.

geraniol A floral, rosy, and sweet-smelling molecule found in a variety of floral scents.

gymnosperm A plant with seeds not enclosed in a protective covering.

ha-ha A sunken fence or wall for a garden or park that is recessed to create a vertical barrier on one side and preserves an uninterrupted view across the landscape on the other.

headspace analysis A process to capture the volatile constituents of an aromatic object, such as a flower, with a trap, or a system for analysis using gas chromatography.

honey bee A eusocial bee in the genus *Apis* that is an important pollinator of crop plants.

indole A powerful molecule found in white flowers that adds an animalic interest to the fragrance.

ionone A chemical compound used in perfumery. Alpha and beta forms of ionone together contribute to the smell of violets; the α form has a fresh scent, while the β form smells powdery and woody.

isomer One of two or more molecules that have the same chemical formula but different structures and often different fragrances.

jasmonate A group of fragrant compounds that add to the fragrance of jasmines and act in signaling in plants.

keystone species A species of plant or animal that is important in an ecosystem and on which other species may depend.

labellum An enlarged petal that forms a lip on a flower.

lactone A constituent of floral scent that often has a milky or fruity aspect.

leafcutter bee A solitary bee in the family Megachilidae that nests in holes lined with cut leaves and can be an important pollinator.

lilac aldehydes Also called syringa aldehydes, a class of aromatics found in plants, including lilacs, that add floral, fresh, and green to fragrances.

limonene A common terpene that has a lemon note.

linalool A terpene alcohol found in many plants that has a flowery fragrance.

methyl salicylate An aromatic compound found in many plants, including wintergreen, that has a sweet fragrance.

microbe A microorganism such as a bacterium, a yeast or mold, or a virus.

midge A small fly that may act as a pollinator but that may also feed on nectar and blood.

mitti Soil.

moth A nocturnal flying insect in the order Lepidoptera that has scaled wings and often a hairy body.

myrcene A terpene found in fragrant plants and herbs that has a peppery and spicy fragrance.

myristicin A compound found in nutmeg that has a warm, woody, and spicy fragrance and may act as an insect repellent.

natural fragrance ingredient An aromatic compound, such as an essential oil, that is isolated from a plant or animal.

nectar guide A mark or structure on the petals of a plant to guide a pollinator to a nectary.

nectary A gland that secretes nectar and is usually found in the flower but may be found on other plant parts.

ocimene An isomer of myrcene with a green floral scent.

outcrossing The process of breeding with unrelated individuals, usually of the same species.

ovary The structure in a plant containing ovules or seeds that may develop into a fruit after fertilization.

perianth Both calyx and corolla in a flower.

petal A modified leaf that is part of a flower, generally brightly colored, and often attractive to pollinators.

phenethyl alcohol Also called PEA, an important constituent of the fragrance of rose that contributes to the "rosy" scent.

pheromone A chemical that carries a message between individuals usually of the same species.

pinene A common terpene with a fresh and earthy pine scent.

piperine A terpene, the pungent aromatic in black pepper.

pistil A single carpel or group of carpels.

pollen The male reproductive cell in plants, generally small and powdery.

pollination The act of fertilization in plants.

pollinator An animal, often an insect, that carries pollen from the anther of a flower to the stigma of a flower.

pollinator syndrome A general way to describe the relationship between pollinator type and flower characteristics.

rhizome A thick, creeping underground stem that produces new shoots and acts as a storage organ.

rose ketones A group of fragrance chemicals that contribute to the scent of a rose.

rose oxide A constituent of the fragrance of rose and flavor in some fruits.

sabinene A constituent of some spices with a woody, camphoraceous, spicy scent and flavor.

sepal A usually green plant part that supports the petal of a flower in bloom. Sepals may also be colored and petal-like in appearance.

sesquiterpene A large, fragrant molecule often found in woods and resins.

skatole A crystalline compound that is generally found in animal feces but may also appear in some plants, including orange blossoms, and has a floral fragrance at very low concentration.

social bee Any of various bees with a defined type of social structure.

solitary bee A large and diverse group of bees that do not have a defined social structure, in which all females can reproduce, and that include some important crop pollinators.

stamen The male sex organ of a flower, consisting of an anther that produces pollen and a stalk or filament.

stigma The upper portion of the pistil that receives pollen grains.

stingless bee Any of a large group of bees with reduced stingers that are generally tropical, are social, and produce honey.

symmetry The arrangement of a flower, either bilateral, in which identical halves are produced only when the flower is cut along a certain axis, or radial, in which cuts in any direction will produce identical halves.

synthetic fragrance ingredient A fragrance ingredient that is generally created from a petrochemical source.

tepal An indistinguishable component of a flower that may be petal or sepal.

terpene Any of a group of diverse fragrant chemicals produced by plants.

terroir The natural environment of a plant that includes soil, climate, elevation.

theobromine A bitter alkaloid found in chocolate and tea.

top note A perfume note generally composed of small volatile molecules such as citrus and some spices or herbs.

umami A class of taste in foods that describes rich, savory flavor.

vanillin An organic compound that produces the characteristic flavor and fragrance of vanilla.

volatile Also called VOC or volatile organic compound, a chemical that evaporates easily.

wadi A valley or channel that may be flooded in the rainy season but is generally dry.

yeast An ancient single-celled member of the fungus kingdom.

zingerone A component of ginger that provides sharp taste and flavor.

Notes

Chapter 1. The Torchwoods

1. Jean H. Langenheim, *Plant Resins: Chemistry, Evolution, Ecology, and Ethnobotany* (Portland, Ore.: Timber Press, 2003), 23–44.

2. Martin Watt and Wanda Sellar, *Frankincense and Myrrh* (Essex, U.K.: C. W. Daniel 1996), 26–28; Martin Booth, *Cannabis: A History* (New York: Thomas Dunne, 2015), chap. 1, Kindle.

3. Charles Sell, *Understanding Fragrance Chemistry* (Carol Stream, Ill.: Allured, 2008), 293–95.

4. William J. Bernstein, *A Splendid Exchange: How Trade Shaped the World* (New York: Atlantic Monthly Press, 2008), 25–26.

5. Arthur O. Tucker, "Frankincense and Myrrh," *Economic Botany* 40 (1986): 425–33.

6. F. Nigel Hepper, "Arabian and African Frankincense Trees," *Journal of Egyptian Archaeology* 55 (August 1969): 66–72; Mulugeta Mokria et al., "The Frankincense Tree *Boswellia neglecta* Reveals High Potential for Restoration of Woodlands in the Horn of Africa," *Forest Ecology and Management* 385 (2017): 16–24.

7. Mulugeta Lemenih and Habtemariam Kassa, *Management Guide for Sustainable Production of Frankincense: A Manual for Extension Workers and Companies Managing Dry Forests for Resin Production and Marketing* (Bogor, Indonesia: Center for International Forestry Research, 2011).

8. Marcello Tardelli and Mauro Raffaelli, "Some Aspects of the Vegetation of Dhofar (Southern Oman)," *Bocconea* 19 (2006): 109–12.

9. Trygve Harris, "About Our Frankincense," Enfleurage Middle East, https://enfleurage.me/about-our-frankincense. Descriptions of frankincense in Oman with excellent photos of the trees in their native habitat.

10. Renata G. Tatomir, "To Cause 'to Make Divine' through Smoke: Ancient Egyptian Incense and Perfume: An Inter- and Transdisciplinary Re-Evaluation of

Aromatic Biotic Materials Used by the Ancient Egyptians," in *Studies in Honour of Professor Alexandru Barnea*, ed. Romeo Cîrjan and Carol Căpiță Muzeul Brăilei Adriana Panaite (Brăila, Romania: Muzeul Brăilei "Carol I"—Editura Istros, 2016).

11. William J. Bernstein, *A Splendid Exchange: How Trade Shaped the World* (New York: Atlantic Monthly Press, 2008), 25–26.

12. Jacke Phillips, "Punt and Aksum: Egypt and the Horn of Africa," *Journal of African History* 38 (1997): 423–57.

13. Jan Retsö, "The Domestication of the Camel and the Establishment of the Frankincense Road from South Arabia," *Orientalia Suecana* 40 (1991): 187–219.

14. Ryan J. Case, Arthur O. Tucker, Michael J. Maciarello, and Kraig A. Wheeler, "Chemistry and Ethnobotany of Commercial Incense Copals, Copal Blanco, Copal Oro, and Copal Negro, of North America," *Economic Botany* 57 (2003): 189–202.

15. Giulia Gigliarelli, Judith X. Becerra, Massimo Cirini, and Maria Carla Marcotullio, "Chemical Composition and Biological Activities of Fragrant Mexican Copal (*Bursera* spp.)," *Molecules* 20 (2015): 22383–94.

16. Philip H. Evans, Judith X. Becerra, D. Lawrence Venable, and William S. Bowers, "Chemical Analysis of Squirt-Gun Defense in *Bursera* and Counterdefense by Chrysomelid Beetles," *Journal of Chemical Ecology* 26 (2000): 745–54.

17. Ryan C. Lynch et al., "Genomic and Chemical Diversity in Cannabis," *Critical Reviews in Plant Sciences* 35 (2016): 349–63.

18. Michael Pollan, *The Botany of Desire: A Plant's-Eye View of the World* (New York: Random House, 2001), 111–80.

19. Christelle M. André, Jean-François Hausman, and Gea Guerriero, "*Cannabis sativa*: The Plant of the Thousand and One Molecules," *Frontiers in Plant Science* 7 (2016), 1–17; Ethan B. Russo, "Taming THC: Potential Cannabis Synergy and Phytocannabinoid-Terpenoid Entourage Effects," *British Journal of Pharmacology* 163 (2011): 1344–64.

Chapter 2. Fragrant Woods

1. Arlene López-Sampson and Tony Page, "History of Use and Trade of Agarwood," *Economic Botany* 72 (2018): 107–29.

2. Regula Naef, "The Volatile and Semi-Volatile Constituents of Agarwood, the Infected Heartwood of *Aquilaria* Species: A Review," *Flavour and Fragrance Journal* 26 (2011): 73–87.

3. Juan Liu et al., "Agarwood Wound Locations Provide Insight into the Association between Fungal Diversity and Volatile Compounds in *Aquilaria sinensis*," *Royal Society Open Science* 6 (2019): 190211; Putra Desa Azren, Shiou Yih Lee, Diana

Emang, and Rozi Mohamed, "History and Perspectives of Induction Technology for Agarwood Production from Cultivated *Aquilaria* in Asia: A Review," *Journal of Forestry Research* 30 (2018): 1–11; Gao Chen, Changqiu Liu, and Weibang Sun, "Pollination and Seed Dispersal of *Aquilaria sinensis* (Lour.) Gilg (Thymelaeaceae): An Economic Plant Species with Extremely Small Populations in China," *Plant Diversity* 38 (2016): 227–32.

4. P. Saikia and M. L. Khan, "Ecological Features of Cultivated Stands of *Aquilaria malaccensis* Lam. (Thymelaeaceae), a Vulnerable Tropical Tree Species in Assamese Homegardens," *International Journal of Forestry Research* 2014 (2014): 1–16; Subhan C. Nath and Nabin Saikia, "Indigenous Knowledge on Utility and Utilitarian Aspects of *Aquilaria malaccensis* Lamk. in Northeast India," *Indian Journal of Traditional Knowledge* 1 (October 2002): 47–58.

5. D. G. Donovan, and R. K. Puri, "Learning from Traditional Knowledge of Non-Timber Forest Products: Penan Benalui and the Autecology of *Aquilaria* in Indonesian Borneo," *Ecology and Society* 9 (2004): 3.

6. Kikiyo Morita, *The Book of Incense: Enjoying the Traditional Art of Japanese Scents* (Tokyo: Kodansha International, 1992), chaps. 2–4; David Howes, "Hearing Scents, Tasting Sights: Toward a Cross-Cultural Multimodal Theory of Aesthetics," in *Art and the Senses,* ed. David Melcher and Francesca Bacci (Oxford: Oxford University Press, 2011), 172–74.

7. Rozi Mohamed and Shiou Yih Lee, "Keeping up Appearances: Agarwood Grades and Quality," in *Agarwood,* ed. Rozi Mohamed (Singapore: Springer Science + Business Media, 2016), 149–67; Pearlin Shabna Naziz, Runima Das, and Supriyo Sen, "The Scent of Stress: Evidence from the Unique Fragrance of Agarwood," *Frontiers in Plant Science* 10 (2019): 840.

8. A. N. Arun Kumar, Geeta Joshi, and H. Y. Mohan Ram, "Sandalwood: History, Uses, Present Status and the Future," *Current Science* 103 (2012): 1408–16.

9. Danica T. Harbaugh and Bruce G. Baldwin, "Phylogeny and Biogeography of the Sandalwoods (*Santalum,* Santalaceae): Repeated Dispersals Throughout the Pacific," *American Journal of Botany* 94 (2007): 1028–40.

10. Harbaugh and Baldwin, "Phylogeny," 1036.

11. B. Dhanya, Syam Viswanath, and Seema Purushothman, "Sandal (*Santalum album* L.) Conservation in Southern India: A Review of Policies and Their Impacts," *Journal of Tropical Agriculture* 48 (2010): 1–10.

12. Kushan U. Tennakoon and Duncan D. Cameron, "The Anatomy of *Santalum album* (Sandalwood) Haustoria," *Canadian Journal of Botany* 84 (2006): 1608–16; P. Balasubramanian, R. Aruna, C. Anbarasu, and E. Santhoshkumar, "Avian

Frugivory and Seed Dispersal of Indian Sandalwood *Santalum album* in Tamil Nadu, India," *Journal of Threatened Taxa* 3 (2011): 1775–77.

13. Mark Merlin and Dan VanRavenswaay, "History of Human Impact on the Genus *Santalum* in Hawai'i," in *Proceedings of the Symposium on Sandalwood in the Pacific, April 9–11, 1990, Honolulu, Hawaii,* USDA Forest Service General Technical Report PSW-122 (Berkeley, Calif.: Pacific Southwest Research Station, 1990), 46–60.

14. Pamela Statham, "The Sandalwood Industry in Australia: A History," in *Proceedings of the Symposium on Sandalwood in the Pacific, April 9–11, 1990, Honolulu, Hawaii,* USDA Forest Service General Technical Report PSW-122 (Berkeley, Calif.: Pacific Southwest Research Station, 1990), 26–38.

15. Jyoti Marwah, "Research Report for Historical Study of Attars and Essence Making in Kannauj" (Navi Mumbai, India: University of Mumbai, 2012–2014).

16. Günther Ohloff, Wilhelm Pickenhagen, and Philip Kraft, *Scent and Chemistry: The Molecular World of Odors* (Zurich: Wiley-VCH, 2012), 39.

Spices

1. Peter Frankopan, *The Silk Roads: A New History of the World* (New York: Alfred A. Knopf, 2015): xiv–xvii.

2. J. M. Haigh, "The British Dispensatory, 1747," *South African Medical Journal* 48 (1974): 2042–44.

Chapter 3. Spices of the Western Ghats

1. C. Elouard et al., "Monitoring the Structure and Dynamics of a Dense Moist Evergreen Forest in the Western Ghats (Kodagu District, Karnataka, India)," *Tropical Ecology* 38 (1997): 193–214.

2. Marjorie Shaffer, *Pepper: A History of the World's Most Influential Spice* (New York: Thomas Dunne Books, St. Martin's Press, 2013), chap. 2, Kindle.

3. Fenglin Gu, Feifei Huang, Guiping Wu, and Hongying Zhu, "Contribution of Polyphenol Oxidation, Chlorophyll and Vitamin C Degradation to the Blackening of *Piper nigrum* L.," *Molecules* 23 (2018): 370–83; K. A. Buckle, M. Rathnawthie, and J. J. Brophy, "Compositional Differences of Black, Green and White Pepper (*Piper nigrum* L.) Oil from Three Cultivars," *Journal of Food Technology* 20 (1985): 599–613.

4. William J. Bernstein, *A Splendid Exchange: How Trade Shaped the World* (New York: Atlantic Monthly Press, 2008), 99–103.

5. K. G. Sajeeth Kumar, S. Narayanan, V. Udayabhaskaran, and N. K. Thulaseedharan, "Clinical and Epidemiologic Profile and Predictors of Outcome of Poison-

ous Snake Bites—an Analysis of 1,500 Cases from a Tertiary Care Center in Malabar, North Kerala, India," *International Journal of General Medicine* 11 (2018): 209–16.

6. Sebastián Montoya-Bustamante, Vladimir Rojas-Díaz, and Alba Marina Torres-González, "Interactions between Frugivorous Bats (Phyllostomidae) and *Piper tuberculatum* (Piperaceae) in a Tropical Dry Forest Remnant in Valle del Cauca, Colombia," *Revista de Biología Tropical* 64 (2016): 701–13.

7. W. John Kress and Chelsea D. Specht, "Between Cancer and Capricorn: Phylogeny, Evolution and Ecology of the Primarily Tropical Zingiberales," *Biologiske Skrifter* 55 (2005): 459–78.

8. P. N. Ravindran and K. N. Babu, eds., *Ginger: The Genus* Zingiber (Boca Raton, Fla.: CRC Press, 2005), 552.

9. Giby Kuriakose, Palatty Allesh Sinu, and K. R. Phivanna, "Domestication of Cardamom (*Elettaria cardamomum*) in Western Ghats, India: Divergence in Productive Traits and a Shift in Major Pollinators," *Annals of Botany* 103 (2009): 727–33; Stuart Farrimond, *The Science of Spice: Understand Flavor Connections and Revolutionize Your Cooking* (New York: DK, 2018), 134–35.

10. Margaret Mayfield and Vasuki V. Belavadi, "Cardamom in the Western Ghats: Bloom Sequences Keep Pollinators in Fields," in *Global Action on Pollination Services for Sustainable Agriculture, Tools for Conservation and Use of Pollination Services: Initial Survey of Good Pollination Practices* (Rome: FAO, 2008), 69–84.

11. Pei Chen, Jianghao Sun, and Paul Ford, "Differentiation of the Four Major Species of Cinnamons (*C. burmannii, C. verum, C. cassia,* and *C. loureiroi*) Using a Flow Injection Mass Spectrometric (FIMS) Fingerprinting Method," *Journal of Agricultural and Food Chemistry* 62 (2014): 2516–21; Gopal R. Mallavarapu and B. R. Rajeswara Rao, "Chemical Constituents and Uses of *Cinnamomum zeylanicum* Blume," in *Aromatic Plants from Asia: Their Chemistry and Application in Food and Therapy,* ed. Leopold Jirovetz, Nguyen Xuân Dung, and V. K. Varshney (Dehradun, India: Har Krishan Bhalla and Sons, 2007), 49–75.

12. Jack Turner, *Spice: The History of a Temptation* (New York: Random House, 2008), 145–82.

Chapter 4. The Spice Islands

1. "The Historic and Marine Landscape of the Banda Islands," UNESCO, January 30, 2015, https://whc.unesco.org/en/tentativelists/6065/.

2. T. R. van Andel, J. Mazumdar, E. N. T. Barth, and J. F. Veldkamp, "Possible Rumphius Specimens Detected in Paul Hermann's Ceylon Herbarium (1672–1679)

in Leiden, The Netherlands," *Blumea—Biodiversity, Evolution and Biogeography of Plants* 63 (2018): 11–19.

3. Manju V. Sharma and Joseph E. Armstrong, "Pollination of *Myristica* and Other Nutmegs in Natural Populations," *Tropical Conservation Science* 6 (2013): 595–607.

4. Diego Francisco Cortés-Rojas, Cláudia R. F. de Souza, and Wanderley Pereira Oliveira, "Clove (*Syzygium aromaticum*): A Precious Spice," *Asian Pacific Journal of Tropical Biomedicine* 4 (2014): 90–96.

5. Stuart Farrimond, *The Science of Spice: Understand Flavor Connections and Revolutionize Your Cooking* (New York: DK, 2018), 12–15.

Chapter 5. Saffron, Vanilla, and Chocolate

1. Maria Grilli Caiola and Antonella Canini, "Looking for Saffron's (*Crocus sativus* L.) Parents," in *Saffron*, ed. Amjad M. Husaini (Ikenobe, Japan: Global Science Books, 2010), 1–14.

2. Juno McKee and A. J. Richards, "Effect of Flower Structure and Flower Colour on Intrafloral Warming and Pollen Germination and Pollen-Tube Growth in Winter Flowering *Crocus* L. (Iridaceae)," *Botanical Journal of the Linnean Society* 128 (1998): 369–84; Casper J. van der Kooi, Peter G. Kevan, and Matthew H. Koski, "The Thermal Ecology of Flowers," *Annals of Botany* 124 (2019): 343–53.

3. Angela Rubio et al., "Cytosolic and Plastoglobule-Targeted Carotenoid Dioxygenases from *Crocus sativus* Are Both Involved in β-Ionone Release," *Journal of Biological Chemistry* 283 (2008): 24816–25.

4. Victoria Finlay, *Color: A Natural History of the Palette* (New York: Random House, 2004), 202–44; Corine Schleif, "Medieval Memorials: Sights and Sounds Embodied; Feelings, Fragrances and Flavors Re-membered," *Senses and Society* 5 (2010): 73–92.

5. J. C. Motamayor et al., "Cacao Domestication I: The Origin of the Cacao Cultivated by the Mayas," *Heredity* 89 (2002): 380–86.

6. Luis D. Gómez P., "*Vanilla planifolia,* the First Mesoamerican Orchid Illustrated, and Notes on the de la Cruz-Badiano Codex," *Lankesteriana* 8 (2008): 81–88.

7. Gigant Rodolphe, Séverine Bory, Michel Grisoni, and Pascale Besse, "Biodiversity and Evolution in the *Vanilla* Genus," in *The Dynamical Processes of Biodiversity: Case Studies of Evolution and Spatial Distribution,* ed. Oscar Grillo (InTechOpen, 2011), doi: 10.5772/24567.

8. R. Kahane et al., "Bourbon Vanilla: Natural Flavour with a Future," *Chronica Horticulturae* 48 (2008): 23–29; Nethaji J. Gallage et al., "The Intracellular Localization

of the Vanillin Biosynthetic Machinery in Pods of *Vanilla planifolia*," *Plant and Cell Physiology* 59 (2018): 304–18.

9. Corine Cochennec and Corinne Duffy, "Authentication of a Flavoring Substance: The Vanillin Case," *Perfumer and Flavorist* 44 (2019): 30–43.

10. Kimberley J. Hockings, Gen Yamakoshi, and Tetsuro Matsuzawa, "Dispersal of a Human-Cultivated Crop by Wild Chimpanzees (*Pan troglodytes verus*) in a Forest–Farm Matrix," *International Journal of Primatology* 38 (2016): 172–93; Maarten van Zonneveld et al., "Human Diets Drive Range Expansion of Megafauna-Dispersed Fruit Species," *PNAS* 115 (2018): 3326–31.

11. Kofi Frimpong-Anin, Michael K. Adjaloo, Peter K. Kwapong, and William Oduro, "Structure and Stability of Cocoa Flowers and Their Response to Pollination," *Journal of Botany* 2014 (2014): 513623.

12. Eric A. Frimpong, Barbara Gemmill-Herren, Ian Gordon, and Peter K. Kwapong, "Dynamics of Insect Pollinators as Influenced by Cocoa Production Systems in Ghana," *Journal of Insect Pollination* 5 (2011): 74–80.

13. Crinan Jarrett et al., "Moult of Overwintering Wood Warblers *Phylloscopus sibilatrix* in an Annual-Cycle Perspective," *Journal of Ornithology* 162 (2021): 645–53.

14. Veronika Barišić et al., "The Chemistry Behind Chocolate Production," *Molecules* 24 (2019): 3163; Van Thi Thuy Ho, Jian Zhao, and Graham Fleet, "Yeasts Are Essential for Cocoa Bean Fermentation," *International Journal of Food Microbiology* 174 (2014): 72–87.

15. Efraín M. Castro-Alayo et al., "Formation of Aromatic Compounds Precursors During Fermentation of Criollo and Forastero Cocoa," *Heliyon* 5 (2019): e01157.

Scented Gardens and Aromatic Herbs

1. W. E. Friedman, "The Meaning of Darwin's 'Abominable Mystery,' " *American Journal of Botany* 96 (2009): 5–21; L. Eiseley, *The Immense Journey* (New York: Vintage Books, 1957), 61–76.

2. Stephen Buchmann, *The Reason for Flowers: Their History, Culture, Biology, and How They Change Our Lives* (New York: Scribner, 2015), 44–61.

3. Chelsea D. Specht and Madelaine E. Bartlett, "Flower Evolution: The Origin and Subsequent Diversification of the Angiosperm Flower," *Annual Review of Ecology, Evolution, and Systematics* 40 (2009): 217–43.

4. Dani Nadel et al., "Earliest Floral Grave Lining from 13,700–11,700-Y-Old Natufian Burials at Raqefet Cave, Mt. Carmel, Israel," *PNAS* 110 (2013): 11774–78.

Chapter 6. Gardens

1. S. Yoshi Maezumi et al., "The Legacy of 4,500 Years of Polyculture Agroforestry in the Eastern Amazon," *Nature Plants* 4 (2018): 540–47.

2. Cynthia Barnett, *Rain: A Natural and Cultural History* (New York: Crown, 2015), 210–26; Saba Tabassum, S. Asif, and A. Naqvi, "Traditional Method of Making Attar in Kannauj," *International Journal of Interdisciplinary Research in Science Society and Culture (IJIRSSC)* 2 (2016): 71–80.

3. L. Mahmoudi Farahani, "Persian Gardens: Meanings, Symbolism, and Design," *Landscape Online* 46 (2016): 1–19; Negar Sanaan Bensi, "The Qanat System: A Reflection on the Heritage of the Extraction of Hidden Waters," in *Adaptive Strategies for Water Heritage: Past, Present and Future,* ed. Carola Hein (Basel: Springer International, 2020), 40–57.

4. Phil L. Crossley, "Just Beyond the Eye: Floating Gardens in Aztec Mexico," *Historical Geography* 32 (2004): 111–35.

5. Wikimedia Contributors, "Gardens of Bomarzo," Wikipedia, https://en.wikipedia.org/wiki/Gardens_of_Bomarzo.

6. Christopher Thacker, *The History of Gardens* (Kent, U.K.: Croom Helm 1979), 263–65.

7. Carolyn Fry, *The Plant Hunters: The Adventures of the World's Greatest Botanical Explorers* (London: Andre Deutsch, 2017).

8. Daniel Stone, *The Food Explorer: The True Adventures of the Globe-Trotting Botanist Who Transformed What America Eats* (New York: Dutton, 2019), 39.

9. Greg Grant and William C. Welch, *The Rose Rustlers* (College Station: Texas A&M University Press, 2017).

Chapter 7. Fragrant Flowers and Aromatic Herbs

1. Alice Walker, *In Search of Our Mothers' Gardens: Prose,* reprint ed. (Boston: Mariner Books, 1983); Dianne D. Glave, "Rural African American Women, Gardening, and Progressive Reform in the South," in *"To Love the Wind and the Rain": African Americans and Environmental History,* ed. Dianne D. Glave and Mark Stoll (Pittsburgh, Pa.: University of Pittsburgh Press, 2006), 37–41.

2. Pat Willmer, *Pollination and Floral Ecology* (Princeton, N.J.: Princeton University Press, 2011), 261–63, 434–35.

3. Siti-Munirah Mat Yunoh, "Notes on a Ten-Perigoned *Rafflesia azlanii* from the Royal Belum State Park, Perak, Peninsular Malaysia," *Malayan Nature Journal* 72 (2020): 11–17.

4. Robert A. Raguso, "Wake Up and Smell the Roses: The Ecology and Evolution of Floral Scent," *Annual Review of Ecology, Evolution, and Systematics* 39 (2008): 549–69.

5. John Paul Cunningham, Chris J. Moore, Myron P. Zalucki, and Bronwen W. Cribb, "Insect Odour Perception: Recognition of Odour Components by Flower Foraging Moths," *Proceedings of the Royal Society B* 273 (2006): 2035–40.

6. Alison Abbott, "Plant Biology: Growth Industry," *Nature* 468 (2010): 886–88; Pavan Kumar, Sagar S. Pandit, Anke Steppuhn, and Ian T. Baldwin, "Natural History-Driven, Plant-Mediated RNAi-Based Study Reveals *CYP6B46*'s Role in a Nicotine-Mediated Antipredator Herbivore Defense," *PNAS* 111 (2014): 1245–52.

7. Danny Kessler, Celia Diezel, and Ian T. Baldwin, "Changing Pollinators as a Means of Escaping Herbivores," *Current Biology* 20 (2010): 237–42.

8. Anne Charlton, "Medicinal Uses of Tobacco in History," *Journal of the Royal Society of Medicine* 97 (2004): 292–96; Sterling Haynes, "Tobacco Smoke Enemas," *BC Medical Journal* 54 (2012): 496–97; Christine Makosky Daley et al., " 'Tobacco Has a Purpose, Not Just a Past': Feasibility of Developing a Culturally Appropriate Smoking Cessation Program for a Pan-Tribal Native Nation," *Medical Anthropology Quarterly* 20 (2006): 421–40.

9. E. A. D. Mitchell et al., "A Worldwide Survey of Neonicotinoids in Honey," *Science* 358 (2017): 109–11; Thomas James Wood and Dave Goulson, "The Environmental Risks of Neonicotinoid Pesticides: A Review of the Evidence Post 2013," *Environmental Science and Pollution Research* 24 (2017): 17285–325; Ken Tan et al., "Imidacloprid Alters Foraging and Decreases Bee Avoidance of Predators," *PLoS ONE* 9 (2014): e102725.

10. Philip W. Rundel et al., "Mediterranean Biomes: Evolution of Their Vegetation, Floras, and Climate," *Annual Review of Ecology, Evolution, and Systematics* 47 (2016): 383–407.

11. Ben P. Miller and Kingsley W. Dixon, "Plants and Fire in Kwongan Vegetation," in *Plant Life on the Sandplains in Southwest Australia: A Global Biodiversity Hotspot*, ed. Hans Lambers (Perth: University of Western Australia Publishing, 2014), 147–70; Şerban Proches et al., "An Overview of the Cape Geophytes," *Biological Journal of the Linnean Society* 87 (2006): 27–43; David Barraclough and Rob Slotow, "The South African Keystone Pollinator *Moegistorhynchus longirostris* (Wiedemann, 1819) (Diptera: Nemestrinidae): Notes on Biology, Biogeography and Proboscis Length Variation," *African Invertebrates* 51 (2010): 397–403.

12. Robert A. Raguso and Eran Pichersky, "New Perspectives in Pollination Biology: Floral Fragrances. A Day in the Life of a Linalool Molecule: Chemical Communication

in a Plant-Pollinator System. Part 1: Linalool Biosynthesis in Flowering Plants," *Plant Species Biology* 14 (1999): 95–120.

13. Maria Lis-Balchin, ed., *Lavender: The Genus "Lavandula"* (London: Taylor and Francis, 2002), 45.

14. Yann Guitton et al., "Differential Accumulation of Volatile Terpene and Terpene Synthase mRNAs During Lavender (*Lavandula angustifolia* and *L. x intermedia*) Inflorescence Development," *Physiologia Plantarum* 138 (2010): 150–63.

15. Carlos M. Herrera, "Daily Patterns of Pollinator Activity, Differential Pollinating Effectiveness, and Floral Resource Availability, in a Summer-Flowering Mediterranean Shrub," *Oikos* 58 (1990): 277–88.

Chapter 8. Roses

1. Natalia Dudareva and Eran Pichersky, "Biochemical and Molecular Genetic Aspects of Floral Scents," *Plant Physiology* 122 (2000): 627–34.

2. Constance Classen, *Worlds of Sense: Exploring the Senses in History and Across Cultures* (London: Routledge, 1993), chap. 1; Alain Corbin, *The Foul and the Fragrant: Odour and the Social Imagination* (London: Papermac/Macmillan, 1986), 89–100.

3. Charles Quest-Ritson and Brigid Quest-Ritson, *The American Rose Society Encyclopedia of Roses: The Definitive A–Z Guide to Roses* (London: DK Adult, 2003).

4. Peter E. Kukielski with Charles Phillips, *Rosa: The Story of the Rose* (New Haven: Yale University Press, 2021), 164–70.

5. A. S. Shawl and Robert Adams, "Rose Oil in Kashmiri India," *Perfumer & Flavorist* 34 (2009): 22–25.

6. P. I. Orozov, *The Rose—Its History* (Kazanlak, Bulgaria: Petko Iv. Orozoff et Fils, n.d.).

7. Gabriel Scalliet et al., "Scent Evolution in Chinese Roses," *PNAS* 105 (2008): 5927–32; Jean-Claude Caissard et al., "Chemical and Histochemical Analysis of 'Quatre Saisons Blanc Mousseux,' a Moss Rose of the *Rosa x damascena* Group," *Annals of Botany* 97 (2006): 231–38.

8. P. G. Kevan, "Pollination in Roses," in *Reference Module in Life Sciences* (Elsevier, 2017), https://doi.org/10.1016/B978-0-12-809633-8.05070-6.

9. Robert Dressler, *The Orchids: Natural History and Classification* (Cambridge, Mass.: Harvard University Press, 1981), 6–9, 140–41.

10. Salvatore Cozzolino and Alex Widmer, "Orchid Diversity: An Evolutionary Consequence of Deception?," *Trends in Ecology and Evolution* 20 (2005): 487–94; Stephen L.

Buchmann and Gary Paul Nabhan, *The Forgotten Pollinators* (Washington, D.C.: Island Press/Shearwater Books, 1996), 47–64.

11. Claire Micheneau, Steven D. Johnson, and Michael F. Fay, "Orchid Pollination: From Darwin to the Present Day," *Botanical Journal of the Linnean Society* 161 (2009): 1–19.

Perfumery from Mandarin to Musk

1. Ann Harman, *Harvest to Hydrosol: Distill Your Own Exquisite Hydrosols at Home* (Fruitland, Wash.: IAG Botanics, 2015), 3–7.

2. Ernest Guenther, *The Essential Oils,* vol. 1: *History—Origin in Plants—Production—Analysis* (New York: D. Van Nostrand, 1948), 189–98.

Chapter 9. Humble Beginnings

1. Mandy Aftel, *Fragrant: The Secret Life of Scent* (New York: Riverhead Books, 2014), 79–122; Dan Allosso, *Peppermint Kings: A Rural American History* (New Haven: Yale University Press, 2020).

2. John C. Leffingwell et al., "Clary Sage Production in the Southeastern United States," in *6th International Congress of Essential Oils* (San Francisco, 1974).

3. William M. Ciesla, Non-Wood Forest Products from Conifers (Rome: FAO, 1998); Cassandra Y. Johnson and Josh McDaniel, "Turpentine Negro," in *"To Love the Wind and the Rain": African Americans and Environmental History,* ed. Dianne D. Glave and Mark Stoll (Pittsburgh, Pa.: University of Pittsburgh Press, 2006), 51–62.

4. Susan Trapp and Rodney Croteau, "Defensive Resin Biosynthesis in Conifers," *Annual Review of Plant Biology* 52 (2001): 689–724.

5. Aljos Farjon, "The Kew Review: Conifers of the World," *Kew Bulletin* 73 (2018): 8.

6. Herbert L. Edlin, *Know Your Conifers*, Forestry Commission Booklet No. 15 (London: Her Majesty's Stationery Office, 1966); James E. Eckenwalder, *Conifers of the World: The Complete Reference* (Portland, Ore.: Timber Press, 2009).

7. S. Khuri et al., "Conservation of the *Cedrus libani* Populations in Lebanon: History, Current Status and Experimental Application of Somatic Embryogenesis," *Biodiversity and Conservation* 9 (2000): 1261–73.

Chapter 10. Perfume Notes

1. G. W. Septimus Piesse, *The Art of Perfumery and Method of Obtaining the Odors of Plants* (Philadelphia: Lindsay and Blakiston, 1857), 162–63.

2. "The Skills Related to Perfume in Pays de Grasse: The Cultivation of Perfume Plants, the Knowledge and Processing of Natural Raw Materials, and the Art of Perfume Composition," UNESCO, https://ich.unesco.org/en/RL/the-skills-related-to-perfume-in-pays-de-grasse-the-cultivation-of-perfume-plants-the-knowledge-and-processing-of-natural-raw-materials-and-the-art-of-perfume-composition-01207.

3. J. W. Kesterson, R. Hendrickson, and R. J. Braddock, *Florida Citrus Oils* (Gainesville: University of Florida Agricultural Experiment Stations, 1971), 6–7.

4. Steffen Arctander, *Perfume and Flavor Materials of Natural Origin* (Carol Stream, Ill.: Allured, 1994), 69–70.

5. Gina Maruca et al., "The Fascinating History of Bergamot (*Citrus bergamia* Risso & Poiteau), the Exclusive Essence of Calabria: A Review," *Journal of Environmental Science and Engineering A* 6 (2017): 22–30.

6. Guohong Albert Wu et al., "Genomics of the Origin and Evolution of Citrus," *Nature* 554 (2018): 311–16.

7. Pierre-Jean Hellivan, "Jasmine: Reinventing the 'King of Perfumes,' " *Perfumer & Flavorist* 34 (2009): 42–51.

8. Peter Green and Diana Miller, *The Genus* Jasminum *in Cultivation* (Kew, U.K.: Royal Botanic Gardens, Kew, 2010), 1; A. B. Camps, "Atypical Jasmines in Perfumery," *Perfumer & Flavorist* 34 (2009): 20–26.

9. N. S. Lestari, "Jasmine Flowers in Javanese Mysticism," *International Review of Humanities Studies* 4 (2019): 192–200.

10. International Federation of Essential Oils and Aroma Trades, "Jasmine: An Overview of Its Essential Oils and Sources," *Perfumer & Flavorist,* January 28, 2019, www.perfumerflavorist.com/fragrance/rawmaterials/natural/Jasmine-An-Overview-of-its-Essential-Oils--Sources-504866941.html.

11. Olivier Cresp, Jacques Cavalier, Pierre-Alain Blanc, and Alberto Morillas, "Let There Be Light: 50 Years of Hedione," *Perfumer & Flavorist* 36 (2011): 24–26.

12. Ziqiang Zhu and Richard Napier, "Jasmonate—a Blooming Decade," *Journal of Experimental Botany* 68 (2017): 1299–302; Parvaiz Ahmad et al., "Jasmonates: Multifunctional Roles in Stress Tolerance," *Frontiers in Plant Science* 7 (2016): 1–15.

13. Selena Gimenez-Ibanez and Roberto Solano, "Nuclear Jasmonate and Salicylate Signaling and Crosstalk in Defense Against Pathogens," *Frontiers in Plant Science* 4 (2013): 72.

14. Rodrigo Barba-Gonzalez et al., "Mexican Geophytes I. The Genus *Polianthes,*" *Floriculture and Ornamental Biotechnology* 6 (2012): 122–28.

15. A. J. Beattie, "The Floral Biology of Three Species of *Viola,*" *New Phytologist* 68 (1969): 1187–201; J. P. Bizoux et al., "Ecology and Conservation of Belgian Popula-

tions of *Viola calaminara*, a Metallophyte with a Restricted Geographic Distribution," *Belgian Journal of Botany* 137 (2004): 91–104.

16. Günther Ohloff, Wilhelm Pickenhagen, and Philip Kraft, *Scent and Chemistry: The Molecular World of Odors* (Zurich: Wiley-VCH, 2012), 191–92.

17. Jianxin Fu et al., "The Emission of the Floral Scent of Four *Osmanthus fragrans* Cultivars in Response to Different Temperatures," *Molecules* 22 (2017): 430.

18. Joshua P. Shaw, Sunni J. Taylor, Mary C. Dobson, and Noland H. Martin, "Pollinator Isolation in Louisiana Iris: Legitimacy and Pollen Transfer," *Evolutionary Ecology Research* 18 (2017): 429–41.

19. Kapil Kishor Khadka and Douglas A. James, "Habitat Selection by Endangered Himalayan Musk Deer (*Moschus chrysogaster*) and Impacts of Livestock Grazing in Nepal Himalaya: Implications for Conservation," *Journal for Nature Conservation* 31 (2016): 38–42; Thinley Wangdi et al., "The Distribution, Status and Conservation of the Himalayan Musk Deer *Moschus chrysogaster* in Sakteng Wildlife Sanctuary," *Global Ecology and Conservation* 17 (2019): e00466.

20. D. Mudappa, "Herpestids, Viverrids and Mustelids," in *Mammals of South Asia,* ed. A. J. T. Johnsingh and Nima Manjrekar, vol. 1 (Hyderabad, India: Universities Press, 2012), 471–98.

21. R. Clarke, "The Origin of Ambergris," *Latin American Journal of Aquatic Mammals* 5 (2006): 7–21; Christopher Kemp, *Floating Gold: A Natural (and Unnatural) History of Ambergris* (Chicago: University of Chicago Press, 2012), 13–16.

Chapter 11. Impossible Flowers and Building a Perfume

1. Marc O. Waelti, Paul A. Page, Alex Widmer, and Florian P. Schiestl, "How to Be an Attractive Male: Floral Dimorphism and Attractiveness to Pollinators in a Dioecious Plant," *BMC Evolutionary Biology* 9 (2009): 190; S. Dötterl and A. Jürgens, "Spatial Fragrance Patterns in Flowers of *Silene latifolia*: Lilac Compounds as Olfactory Nectar Guides?," *Plant Systematics and Evolution* 255 (2005): 99–109.

2. Chloé Lahondère et al., "The Olfactory Basis of Orchid Pollination by Mosquitoes," *PNAS* 117 (2020): 708–16.

3. G. W. Septimus Piesse, *The Art of Perfumery and Method of Obtaining the Odors of Plants* (Philadelphia: Lindsay and Blakiston, 1857).

Chapter 12. Scented Worlds

1. Alain Corbin, *The Foul and the Fragrant: Odour and the Social Imagination* (London: Papermac/Macmillan 1986), 176–81.

2. George S. Clark, "An Aroma Chemical Profile: Coumarin," *Perfumer and Flavorist* 20 (1995): 23–34; Robin Wall Kimmerer, *Braiding Sweetgrass: Indigenous Wisdom, Scientific Knowledge, and the Teachings of Plants* (Minneapolis, Minn.: Milkweed Editions, 2013), 156–57.

3. Günther Ohloff, Wilhelm Pickenhagen, and Philip Kraft, *Scent and Chemistry: The Molecular World of Odors* (Zurich: Wiley-VCH, 2012), 7.

4. Anne-Dominique Fortineau, "Chemistry Perfumes Your Daily Life," *Journal of Chemistry Education* 81 (2004): 45–50.

5. Tilar J. Mazzeo, *The Secret of Chanel No. 5: The Intimate History of the World's Most Famous Perfume* (New York: Harper Perennial, 2011), 59–72.

6. Jean-Claude Ellena, *Perfume: The Alchemy of Scent* (New York: Arcade, 2011).

7. Catherine L. Ludlow et al., "Independent Origins of Yeast Associated with Coffee and Cacao Fermentation," *Current Biology* 26 (2016): 965–71.

8. Erick J. Vandamme, "Bioflavours and Fragrances via Fungi and Their Enzymes," *Fungal Diversity* 13 (2003): 153–66.

9. Anya McCoy, *Homemade Perfume: Create Exquisite, Naturally Scented Products to Fill Your Life with Botanical Aromas* (Salem, Mass.: Page Street, 2018).

10. Mandy Aftel, *Essence and Alchemy: A Natural History of Perfume* (Layton, Utah: Gibbs Smith, 2001); Harold McGee, *Nose Dive: A Field Guide to the World's Smells* (New York: Penguin, 2020).

Index

Page numbers in italics indicate illustrations.